T0231568

LAND REFORM AND SUSTAINABLE DEVELOPMENT

To my wife and family

and

To the staff and students, past, present and future,
of the School of Surveying, University of East London.

May the next 50 years be as successful as the last half century.

Land Reform and Sustainable Development

Edited by
ROBERT W. DIXON-GOUGH
School of Surveying, University of East London

Routledge
Taylor & Francis Group

LONDON AND NEW YORK

First published 1999 by Ashgate Publishing

Reissued 2018 by Routledge
2 Park Square, Milton Park, Abingdon, Oxon, OX14 4RN
52 Vanderbilt Avenue, New York, NY 10017

Routledge is an imprint of the Taylor & Francis Group, an informa business

Publisher's Note
The publisher has gone to great lengths to ensure the quality of this reprint but points out that some imperfections in the original copies may be apparent.

Disclaimer
The publisher has made every effort to trace copyright holders and welcomes correspondence from those they have been unable to contact.

A Library of Congress record exists under LC control number: 99072652

ISBN 13: 978-1-138-36952-8 (hbk)
ISBN 13: 978-0-429-42860-9 (ebk)

Contents

List of contributors

G. Andonov. Bulgarian Association of Valuers of Real Property.

C. Arnison. Professor of Rural Economy and Land Management, Royal Agricultural College, Cirencester, England.

K. Batanov. Bulgarian Association of Valuers of Real Property.

Dr. A. Brimicombe. Professor in Surveying, School of Surveying, University of East London, England.

Professor Dr. A. van den Brink. Government Service for Land and Water Use, Utrecht *and* Department of Physical Planning and Rural Development, Wageningham Agricultural University, The Netherlands.

Dr. R.K. Bullard. Director of the Land Reform Research Unit, School of Surveying, University of East London, England.

R.W.Dixon-Gough. Member of the Land Reform Research Unit, School of Surveying, University of East London.

D.C. Doughty. Researcher, School of Surveying, Kingston University, England.

Dr. R. Home. Reader in Planning, School of Surveying, University of East London, England.

Dipl.-Ing. E.C. Läpple. Bundenministerium für Ernahrung, Landwirtschaft und Forsten, Bonne, Germany.

Dr. G. Leidig. Secretary to the Council of Administration for the European Faculty of Land Use and Development. Bundesverband Druck E.V., Wiesbaden, Germany.

Professor Dr. Dr. h.c. mult H. Lenk. Dean of the European faculty of Land Use and Development, Institute of Philosophy, University of Karlruhe, Germany.

Ass. Professor Dr. R. Mansberger. Institute of Surveying, Remote Sensing and Land Information, University of Agricultural Sciences, Vienna, Austria.

Dr. J. Nyiri. College for Surveying and Land Management, University of Sopron, Székesfehérfár, Hungary.

Dipl.-Ing. I. Pesl. Surveying and Cadastral Inspectorate, Opava, Czech Republic.

M. Rizov. Bulgarian Association of Valuers of Real Property.

Ir. F.B. Rosman. Faculty of Geodetic Engineering, Real Estate Law and Planning, Delft University of Technology, The Netherlands.

Dipl.-Ing. J. Rydval. Cadastral Offices, Blansko, Czech Republic.

Ass. Professor W. Seher. Institute of Regional Planning and Rural Development, University of Agricultural Sciences, Vienna, Austria.

Dipl.-Ing. J.K.B. Sonnenberg. Director of Land Consolidation, kadastre, Apeldorn, The Netherlands.

O. Univ. Professor Dr. G. Stolitzka. Director of the Institute of Surveying, Remote Sensing and Land Information, University of Agricultural Sciences, Vienna, Austria.

Professor Dr. P. Trappe. Department of Sociology, University of Basle, Switzerland.

Professor Dr. G. Weber. Director of the Institute of Regional Planning and Rural Development, University of Agricultural Sciences, Vienna, Austria.

D. Williams. Researcher, School of Surveying, University of East London, England.

Ir. H.W. de Wolff. Faculty of Geodetic Engineering, Geo-Information and Land Development, Delft University of Technology, The Netherlands.

W.G.O. Zwirner. Reader in European Property, School of Surveying, Kingston University, England.

Preface

Dr. A.De LEEUW

The European Faculty of Land Use and Development (Faculté Européen des Sciences du Foncier; Europäische Fakultät für Bodenordung) was founded in 1980 at Strasbourg. It has a mission to study and to develop interdisciplinary scientific methodology concerned specifically with the problems of land use and development across Europe, including such themes as urbanism, land management and the environment. In addition, its aim is to contribute towards the gaining of a greater degree of understanding of aspects of national legislation, across Europe, through the wide diversity of disciplines of the Faculty. This mission is accomplished by means of education and higher level research within the university sector throughout Europe.

The Faculty is a forum for discussion, comparison and exchanges of ideas and experiences, studies and research across those subjects within it's broad scope. Within the framework of these activities, the Faculty co-operates with universities and research organisations both within Europe and internationally. The dissemination of this research is through the organisation of regular symposiums.

An Administration Council, for whom I have the honour to be its President, directs the Faculty: the official location being in Strasbourg (9, Place Kléber). It operates with the permission of the relevant Fench authorities and is a legal French association. The status of the association dates from the 12th December 1979 and it was registered in the "Registre des associations" of the "Tribunal d'Instance de Strasbourg" on the 17th September 1980. The Faculty is, at present, a group of 36 professors and readers, attached to 30 European universities. Each member is a recognised specialist in an aspect of land use and development.

To date, twenty-seven symposiums (each of a duration of between 2 and 3 days) have been organised, usually twice a year. Each symposium normally includes a technical excursion related to the theme of the symposium. Three symposiums have been organised by myself in Strasbourg, seven in collaboration with the Council of Europe and seventeen in collaboration with the following universities: Granada, Cambridge,

Lublin (Poland), the Royal Agricultural College (Cirencester), Vienna (twice at the Universität für Bodenkultur), Yildiz (Istambul), Zurich (twice at the Eidenössische Technische Hochschule), Delft, Pireus, Kingston (twice), Munich (Technische Universität), Bonn (in collaboration with the European Centre of the University of Waseda, Tokyo), Olsztyn (Poland), the University of East London, Sopron (Hungary), and Portsmouth. Likewise, the study visits have been organised to Graz (Austria) at the Alfred Pikalo Institute.

The location and themes of the last four Symposiums have been:

No. 25 May 1997   University of Sopron (Székesfehévar)
"Land use Policy for Specific Applications"

No. 26 September 1997   University of Portsmouth
"Coastal Zone Management. Partnership approaches to CZM"

No. 27 May 1998   Eidenössische Technische Hochschule, Zurich
"Regional/Spatial Planning in a State of Flux"

No. 28 September 1998   Kingston University
"Land Use Demand for Leisure and Recreation"

It is notable that the activities of the Faculty are carried out without any official financial support and only exists through the grace of some private sponsors.

The choice of the themes for the symposiums must address one or more of the following three criteria; a study of the topical subjects concerned with land use and development, to discuss the problems in an interdisciplinary manner, and to place the work in a European context. Our idea of Europe is not that of the European Union or even the Council of Europe, but one that stretches from the Atlantic to the Urals.

The working method of the Faculty is that a gathering, of up to a maximum of 20 to 25 participants, will present their papers - without interpretation - in one of the three working languages; English, French and German. The papers are discussed and, when possible, the reviewed and edited papers are published. To date, 21 volumes (4 monographs and 17 symposiums) have been published by Peter Lang (Frankfurt am Main, Bern). However, because of financial difficulties, the publication of the individual symposiums was suspended and this new formula of presentation and publication is being adopted.

The Faculty's relationship with other Universities outside Europe should also be pointed out, notably Waseda Univerity of Tokyo and the University of Buenos Aires and Cordoba in Argentina. Thus, the Faculty has also been instrumental in playing a part in the scientific exchanges between Europe and those countries.

This present volume contains the papers presented at the 23rd Symposium organised by the School of Surveying of the University of East London, the theme being "Land Reform and Sustainable Development".

Strasbourg, June 1998

Dr. A. De Leeuw
Président du Consiel d'Administration de la Faculté Européenne

1 Land reform: the key to sustainable development

R.W.DIXON-GOUGH

During October 1995, the University of East London hosted the 21st International Symposium of the European Faculty of Land Use and Development, organised by Dr Richard Bullard. One of the characteristics of the Faculty is that its meetings are hosted by Institutions who define the theme of the symposium. All participants are expected to present and deliver their papers, which are then discussed. This book is based upon the papers delivered and discussed at this symposium. All papers have been extensively re-written and edited to incorporate both the results of the discussions that took place at the symposium and the developments that have subsequently taken place in this, very dynamic field of activity.

The themes of land reform and sustainable development are inexorably linked and combine two aspects that are invasive in many of the chapters contained within this text, those of human and technical resources. This is particularly so in the context of Europe, which is currently undergoing significant changes, both in attitude and practice, concerning the ways in which land is used and managed. Within the European Union (EU), a considerable degree of emphasis is being given to the environment and ecologically-friendly ways of managing the land, whilst in eastern Europe there have been fundamental and significant changes in land management practices leading to whole-scale land and agrarian reform. Dale (1997) comments that

> Of the three factors of production that are said to underpin the creation of wealth – capital, labour and land – it is the land that is least well understood.

The problems of land management involves the definition of a close relationship between land, the ecology and the social conditions, together with an understanding of the economic structures within both rural and urban areas. In the Land Administration Guidelines, recently produced by the United Nations Economic Commission for Europe (UN ECE), land management is described as the process by which the resources of the land are put to the best possible effect. This includes such procedures as farming, mineral extraction, property and estate management and the physical planning of urban and rural regions, all of which are covered within this book. Furthermore, and fundamental to the processes of land management, are the management of land resources, the formation of land use policies, the impacts of the land upon socio-economic conditions, and the monitoring of all activities on the land to ensure its best possible use.

The efficient use of land is frequently related to the prevailing socio-economic conditions of the region. In the 1972 United Nations Conference on the Human Environment, held in Stockholm, Indira Gandhi stated that 'poverty is the biggest polluter'. People in this state frequently subsist in an unsustainable manner by overgrazing, devastating woodland and natural vegetation by collecting and cutting wood for use as fuel, etc.. Conversely, however, overcoming poverty has also increased stress on the environment with a higher *per capita* consumption of water, energy, and the creation of greater stress upon the land through various forms of pollution (Dixon-Gough *et al*, 1998). This comment by Indira Gandhi, was echoed more recently by Litvin (1998) who states that

> Conventional wisdom has it that concern for the environment is a luxury that only the rich can afford; that only people whose basic needs for food and shelter have been met, can start worrying about the state of the health of the planet.

Such, therefore, are the problems of land management, which encompasses the concepts of land reform and sustainable development.

Whilst land reform is very tangible, the concept of sustainability is not and, in common with a great many terms in current use, it has a variety of meanings (Cullingworth & Nadin, 1994). This is very ably demonstrated in this book. Shiva (1992) identifies two very different meanings to the concept of sustainability. The first relates to, what is generally accepted as, the 'real meaning', which relates to the primacy of nature

> Sustaining nature implies maintaining the integrity of nature's processes, cycles and rhythms.

This definition, however, does not address the concept of the sustainable urban development. Shiva's alternative meaning is concerned with 'market sustainability', which is concerned with conserving resources for development purposes and, if they become depleted, finding substitutes.

In England and Wales, the concept of sustainability is referred to in general terms in the form of a planning policy guidance note PPG12, which states

> The Government has made clear of its intention to work towards ensuring that development and growth are sustainable. It will continue to develop policies consistent with the concept of sustainable development. The planning system, and the preparation of development plans in particular, can contribute to the objectives of ensuring that development and growth are sustainable. The sum total of decisions in the planning field, as elsewhere, should not deny future generations the best of today's environment. This should be expressed through the policies adopted in development planning.

Such a choice of words and expressions are sufficiently vague to satisfy most interests (Cullingworth & Nadin, *op cit*).

One of the most fundamental aspects of land reform and sustainable development is the basic philosophy underlying the processes involved. Boyle and Write (1992) make reference to the 'tip of the iceberg' phenomenon, which acknowledges the multi-layered approach to the problem. It also implies that many informal structures, processes and organisational systems exist below the surface of, what might be fundamentally perceived as, a client or dominant organisation. In the first part of this book, both Lenk and Leidig consider the ecological consequences of land reform and sustainable development. This is indeed an aspect that is fundamental to any programme of sustainability and particularly those involved in rural land management. The ecological implications of sustainable development are not, however, solely related to the rural landscape.

In 1992, the specific problems related to the environment were addressed at the United Nations Earth Summit, held in Rio de Janeiro. One of the outcomes of the meeting was a series of Agendas outlining recommended programmes for the 21st century. In particular, Agenda 21 specifically addressed the concept of sustainable development and

3

environmentally acceptable land management. Both local and urban communities are feeling the implications of Agenda 21 of the Rio Summit throughout Britain and Europe. Through the Local Agenda 21 initiative, plans are being formulated throughout Britain to highlight local concerns, to identify issues and to explain what local councils are doing to help tackle those issues (Thamesdown District Council, 1996). It is yet unclear whether the will or the finance will be made available to implement those plans. There is a general unwillingness to pay for reforms and actions that are needed to ensure long term sustainability (Dale *op cit*).

At the time of the 21st International Symposium, an EU Minister's Conference was taking place in Sofia. This called for the development of new policy frameworks to give stability and predictability in order to help farmers and rural land managers to plan for the long term, and to secure environmental benefits. This conference urged that the principles of sustainable development (as defined by the Bruntland Commission) – developments which meet the needs of this generation without compromising the ability of future generations to meet their needs – be incorporated into all decisions from international down to local levels. The EU was called on

> to work towards an integrated rural sustainable policy, incorporating sustainable development principles, and delivered in such a way which sustains regional and local characteristics

It is, therefore, of interest, given the policies identified at the Sofia conference, that Lenk questions whether there is an adequate model that will allow economic thinking and economic models to measure and assess ecological damages. As with any form of ecological damage, there is the problem of attributing responsibility, which is often confused under an umbrella of collective and group actions. Who should pay for such damage and how should the finances be collected? At what point does the damage become apparent and how much latent damage has to be caused before the actual damage can be physically realised, assessed and measured? This problem is illustrated by Lenk in the concepts of the 'prisoner's dilemma' which is transposed into the 'enjoyer's dilemma'.

Leidig also considers the interrelationship between the ecology and the economy. Furthermore, he introduces the concept that any form of land development is essentially a sub-system of regional science. In this form, the concept relates the planning process, in which the land, property and usage

4

are used to create new system structures, to the future economic/ecological functionality of new systems of society. In other words, he is defining the concept of sustainability, which this generation must use to plan for the future. These processes are defined as 'Ecological Land Development'. Leidig investigates the potential of newer processes of research, such as artificial intelligence, artificial life, virtual reality and chaos theory in opening up the possibility of multidisciplinary co-operation and problem solving.

Very much in keeping with the work of Lenk and Leidig, in relating the importance of ecology to land development and sustainability, is the concept of a multidisciplinary approach to rural land development. In the following chapters, van den Brink, Solitzka and Mansberger, and Sonnenberg address this issue. Sonnenberg discusses the legal instruments that serve multi-functional land development and land reform in the Netherlands. This is related to the Land Development Act of 1995, which provides a wide range of legal instruments for the development of rural areas. He considers that this legislation, in particular the instruments relating to land consolidation and land redevelopment, is applicable to most of the problems encountered in rural land development. van den Brink acknowledges that the planning and rural layout in the Netherlands is geared towards sustainable development and continues this theme. It is clear that, in the Netherlands, the policy of the government on spatial planning, the environment and water, agriculture, nature and the landscape are directed towards a policy of sustainable development. The key event marking the introduction of this policy was the publication of the Bruntland Report in 1987 (WCED, 1987), which ensured that sustainable development was at the forefront of attention to everyone concerned with the future of rural areas. Gordillo & Riddell (1996) comment that:

> Recent summits have addressed the core problems of humankind from different perspectives. The Rio summit was a time of reckoning with the fact that the environment – the resource base that we all share, no matter where we reside – is surprisingly fragile and is in dynamic interchange with human activity. After the Rio summit, it was no longer debatable that what we do individually and institutionally has a profound cumulative affect on the environment.

van den Brink cautions the current approaches to sustainable development for both urban and rural areas. He emphasises that it is important that some form of operational concept be identified in the pursuit of sustainable development. His caution is that all too often, the concept of sustainability becomes an empty phrase or gesture, which results in a 'Babel-like confusion of tongues'. In many cases, the lack of a strictly defined content for sustainable development makes it clear that it is not a goal *per se* but a general principle in which ecological, environmental, social and economic developments can be tested.

Dale (*op cit*) considers that the role of land information in both urban and rural land developments is crucial. Poor decisions are often the result of poor land information. Stolitzca and Mansberger suggest that both scientists and politicians need complete and reliable data concerning the natural resources. Such data would allow them to acquire knowledge relating to the possible causes of likely and potential damages to a certain policy or development. They propose the development of a publicly available and legally binding database of natural resources that is similar, in concept, to the Austrian cadastre and land register. This type of database would permit access by both the public and professionals, either by visiting local offices or accessing the data using modern technology, such as the Internet. The establishment of a countrywide database would very expensive, both in terms of human and technical resources. In Austria, however, both the land register and the cadastre have been converted to a digital framework. Both systems would, therefore, be suitable for providing the basic framework for a natural resource database, with the only change being the definition of the land parcels, which would have to be expanded. The finances required to implement such a system and to acquire the relevant data will be provided for by charging for access to the database.

Many types of environmental modelling procedures relating to land reform and sustainable development are dependent upon the availability of good quality spatial data. Jankowski and Stasik (1997) condered that the use of Geographical Information Systems (GIS) is essential for land zoning and spatial decision making, both operations being fundamental to land reform and sustainable development. Brimicombe gives specific emphasis to the role of GIS and environmental modelling to mitigate against geo-hazards in land use planning and development. As an illustration of this concept, it is indicated that the geo-hazards are characterised by a non-random pattern of spatial distribution and can be expected to occur repeatedly within a reasonable period of time. In contrast to many other themes of sustainable development, which espouse the concept of living 'in harmony with nature', Brimicombe suggests that it is a somewhat romantic notion given the cyclic prevalence of geo-hazards in many parts of the world. It is, however, emphasised that the role of spatial information within a GIS is now an established and valuable tool in the planning process. This does not mean that a GIS should be treated as panacea for all ills since a number of constraining issues do arise, notably the relationship between land use planning and sustainable development and upon certain modelling deficiencies within GIS. The basic form of the spatial data used in a GIS cannot hope to define the cultural landscape and social constraints, which are spatially fluid and often have indistinct boundaries (Brimicombe & Yeung, 1995).

Any form of database or spatial information system can be the key to the success of a market economy, by linking ownership directly to property rights. Furthermore, there is a direct relationship between the degree of care lavished on the land and its ownership and, therefore, ownership and property rights must be defined to create more efficient land markets. In most western European countries, efficient land markets are taken for granted, as is the definition of land. However, in Russia for example, buildings are regarded as separate entities to the land upon which they stand. They can be owned without the owner having any title to the land. The development of land markets is taking place in countries in eastern and central Europe. This is, partly in response to the needs of those countries in economic transition but also because of the impact of information technology. Many of the current land reform programmes were inspired initially from a sense of justice rather than by economic factors (Dale, *op cit*). The processes of land privatisation and re-privatisation have tended to create inefficient patterns of land use, with many farmers being in the position of owning several small land parcels, which might be dispersed over a relatively wide area. This is the inevitable relationship between the human and technical elements, coupled with resources, that are invasive throughout all processes of land management. Data concerning land and property exists in greater quantities than ever before, largely thanks to digital technology.

One of the most fundamental stages in the evolution of any land reform programme is that of agrarian reform. History is littered with examples of agrarian reform that have been implemented on an institutional basis and that have failed disastrously, often leaving the land

and the inhabitants in a far poorer state than before the process had started. Examples include the state and co-operative farms of the former USSR, in Africa under relatively homogeneous conditions (the Sahel Zone), and in Mexico, which has a long tradition of agrarian reform. In the majority of cases of institutional agrarian reform, the driving force has been the re-distribution of agricultural areas into larger 'factory' farms. The processes of agrarian reform practised in the former USSR is discussed within this book by Andonov *et al*, Arnison, Nyiri & Dixon-Gough, and Rydval & Pesl. In each of the above discussions, agrarian reform has been driven by political rather than by the economic infrastructure, and each case has resulted in great social changes. Trappe, however, focuses on the social infrastructure related to agrarian reform within an institutional system, that has withstood the test of specific institutional pre-conditions of a regional or state order. These represent a very good example of the 'tip of the iceberg' phenomenon discussed above. Some of the most significant aspects of such reforms are the infrastructural measures, such as the water catchment areas, existing irrigation systems and the pre-conditions for irrigation farming in lowland river valleys. Trappe discusses, amongst others, the Badajoz plan, which has served to illustrate agrarian reform throughout Spain, and has acted as a very useful pilot scheme. In this programme, it is significant that the stages of expansion projected at the outset of the project were fully achieved. These included the economic infrastructure measures, land re-distribution measures, the settlement of new farmers, the co-operative infrastructure, fully irrigated six hectare cultivation, the planning of agro-industries and the establishment of a dairy industry.

For almost fifty years, the countries of eastern and central Europe where separated from the west by the 'iron curtain'. The socialist system lived by its own laws and standards and once the system had collapsed and the borders opened, it became necessary for these countries to adapt themselves to European standards. It is, however, important to reflect that procedures that work successfully in one country may not necessarily work in another. Dale (*op cit*) emphasised that the over-riding concern must be that programmes, currently being developed for former communist countries with western aid, are designed so that the solutions must fit the environment.

Contract agriculture represents the need for orderly and predicatable transactions that characterise contactual relationships that has, in many parts of Europe, necessitated a new emphasis on the regularisation of land tenure agreements. Many countries are undertaking major planned revisions to their cadastral and land registration systems, in the realisation that the market cannot play its role if the property relationships are too ambiguous to enter into secure and enforceable contracts.

In the past, it was conventional wisdom that agriculture should pattern itself on industry. This concept was tested with vigour on the state and co-operative farms of the former USSR. It has become apparent, not only in the former USSR but in other European countries, that once protection, special subsidies, legal constraints, etc. had been removed, there appeared to be few, if any, economies of scale above the size of a unit that could be managed by a family. This is largely due to the density of management costs in relation to labour costs. This, however, does not necessarily mean a return to small independent units. Where agriculture remains competitive, it is often within the framework of associative organisations of agriculturists. These associations exist on both a formal and informal basis to provide the economy of scales necessary for the purchase and sale of commodities.

Today, agriculture must be capable of taking its place within the institutional framework necessary for participation in the liberalised market economy. Thus, there must be investment in rural institutions (cadastre, land registration, conveyancing, mortgaging, law, etc.) as well as in urban ones.

Arnison comments that one of the assumptions most western observers had when, in 1989/90, the Soviet Union ceased to exercise central control over its economy and that of its satellite states, was that urgent land reforms would be necessary. These reforms would primarily address the issue of the ownership and control of the land. It was assumed that 'the people', in those countries annexed at the end of the Second World War, would demand the restitution of land confiscated from them by the State. This reflects the view of Dale (*op cit*), in that justice rather than economic factors essentially inspired the process of land reform. It was felt at the outset that the ownership of land and property was the single and essential key to the formation of private enterprises, which would be needed to fill the vacuum left by the collapse of centralised production and distribution. This 'simple' concept has generated considerable confusion and many difficulties, which have virtually resulted in the collapse of the Russian economy. Arnison cites two apparent difficulties. The first is concerned with the legal instruments, which granted powers to implement the privatisation legislation to an inappropriate ministry with a vested interest in maintaining status quo. This was the result of an over-simplified approach, both politically and juridically, to the process of privatisation. What is actually required is a suite of laws, rather than a single law, which address the interacting and independent factors relating to ownership. These include the compulsory registration of titles and transactions, the principles of taxation of both annual and capital values, and a law covering land use with controls concerning both its change of use and development. These comments are also reflected by Brooks (1994), Csaki, (1994) and Thiesenhusen (1994), who have remarked upon the piecemeal approach to land reform legislation in Russia.

The second of the difficulties inhibiting the progress of land reform is the relative lack of a professional infrastructure of lawyers, accountants, valuers, surveyors, engineers and public administrators, who would have been in a position to advise on the practical nature of land reform legislation at an early stage.

In contrast to the situation in Russia, where the legislation fails to fully address the complicated issues related to economic and social reform, Rydval and Pesl document, in considerable detail, the level and nature of the complicated legal instruments that have been passed into legislation within the Czech Republic. They clearly identify the nature of the step-by-step processes that have resulted in a transition from state control to a planned market economy. These contrast greatly with the legal provisions in Russia, as identified by Arnison. Rydval and Pesl have identified five significant stages involved in the process of economic reform (institutional, monetary and real) that are fundamental to the entire concept of system reform. These are; denationalisation and privatisation of state ownership, price liberalisation, foreign trade liberalisation and foreign investment, tax reform, and the reform of finance and currency regulations. These basic steps in system reform are mutually conditional and essential in the process and implementation of land reform. This chapter covers three important elements; the legal framework for land restitution and privatisation, the development and evolution of the Czech land cadastre, and the processes of land consolidation and valuation. In the rapid evolution of the processes of land reform in the Czech Republic, the authors emphasise the enormous political pressures that the system has been under to produce results quickly. This has resulted in, for example, the land cadastre being exposed to many political, economic and social pressures since 1990. They also reveal that there had been tendencies to simplify   the complicated cadastral processes, at the expense of both credibility and reliability. This was largely overcome by strictly enforcing the legislation and by adhering to the basic technical principles that are essential for the continued and sustainable maintenance of the cadastre.

Nyiri and Dixon-Gough, who specifically address the concepts of rural planning and land privatisation in Hungary, continue the themes developed by Rydval and Pesl in a more specialised manner. The authors identify four important considerations related to land policies that were developed during the early 1990s. These were; an acknowledgement by the government that a private/real estate market was an essential requirement for the development of a market economy, the development (within a reasonable period of time) of the legislation and executive mechanisms for a viable private/real estate market for the whole economy, the application of the principles of private enterprise to all organisations involved in land registration and, finally, that the costs of the land registration process be funded by the users and be, in general terms, self-financing. One great advantage Hungary had over many of the former communist countries, was that a basic operational structure had been developed in the early 1970s. However, owing to a low priority of ownership and changes in ownership during that period, only land and property data had been computerised. During the latter part of the 1980s, there was an increased level of interest in moving towards a market-orientated economy which, in turn, increased the level of interest in the development of a country-wide computerised system for land registration and cadastral mapping. To illustrate these concepts, the authors present examples of the current status of the established property structure in Hungary, together with the present legal and financial background and processes of land consolidation. These changes have placed considerable pressure upon the system of land registration since all changes should be surveyed and registered, and the land register brought up-to-date. In order to cope with the rate of such changes, a multi-purpose land and property register, together with a digital cadastre, is essential for an effective transition from state-owned land to a market-driven economy. Such a system, TAKAROS, has been developed to cope with these changes. This will eventually be developed as a parcel-based, regional Land Information System (LIS) with GIS capabilities. It will also be capable of being operated between land offices and provide remote access for municipalities, notaries and lawyers. This will permit users, in particular the district offices, to integrate the land registration records within a maintained, topologically-structured parcel fabric.

Any LIS or GIS, such as TAKAROS, requires good quality, accurate data to permit it to operate in an efficient manner, yet the level of funding for cadastral mapping in Hungary was low. To overcome this problem, a National Cadastral Programme was launched as an independent, self-financing government agency. This has been made possible by loans from the German Federal Bank and the Bavarian State Government to promote the renewal of the old Hungarian cadastral maps, with particular emphasis being given to the settlements affected by the TAMA land consolidation project. The final part of the chapter by Nyiri and Dixon-Gough outline the processes of land consolidation in Hungary and provide examples to illustrate the operation of the TAMA project.

The problems of land reform in former communist countries is also discussed in the two following chapters by Andonov and Risov and Andonov *et al*. Andonov *et al* outline the complicated pattern of land tenure in Bulgaria during the country's turbulent political history through the past one hundred years. The authors trace the origins and evolution of modern land tenure systems in the region of South Dobrudja and link those to the region's political history. The legislation, relating to ownership and the use of farming land, that was passed in 1992 essentially addressed the issues of land restitution since it addressed the ownership rights of the original owners at the time of the formation of the co-operative farms. In general, however, the law has created a great deal of confusion since the lack of title proofs, the quantity of them and the contradictory nature of many of them have caused significant delays in the physical process of land reform schedules throughout Bulgaria. The authors have also identified one feature not identified in previous chapters dealing with the process of land reform in the former communist countries, that of the leasing of privately owned land. In Bulgaria, private farms are commonly in excess of 300 ha and in some areas they might exceed 20,000 ha. A further factor is that the majority of the restored land ownership rights are to those either too old to farm or already dead. This is complicated by the lack of an operating land market, despite some attempts at a few preliminary sales, and probably accounts for the substantial number of large farms comprised essentially of leased land.

Andonov and Risov discuss the relationship between land reform and the land market in the following chapter. Whilst recognising that the function of land reform is primarily directed towards the establishment of a free market, they recognise the difficulties faced by the emerging market economy in Bulgaria. The authors have conducted a detailed analysis of the current status of land reform and have identified a number of problems. One of the most important, is related to the character of the documentation and of differing degrees of importance relating to the processes of former land ownership. They also point to the pressures (also mentioned by Rydval and Pesl) exerted upon the specialists in the Municipal Land Commission, by both executive and political powers as individuals, such as Members of Parliament. In other cases, problems arise from the incompetent interpretation of the law, together with subjective and unrealistic recommendations, which are increasingly confusing the contractors. Other factors cited as influencing land reform are unsatisfactory staff and technical equipment and the strongly influencing political partialities at the various levels throughout Bulgaria. With respect to the land market, it is possible (once ownership has been restored and the necessary documents received) for any owner to sell their land. In practice, however, there is still a very limited land market and the majority of land available for disposal, both from the private sector and the state, is leased over periods of between and three years. This can lead, in some instances, to the existence of leased farms of up to 90,000 ha in size, although the average size of leased farm is between 300 and 800 ha.

The following two chapters by Rosman and Weber move towards a discussion of the general principles of sustainability within the context of spatial development and planning. Rosman considers the issues of sustainable development within the educational curriculum of the Faculty of Geodetic Engineering, Real Estate Law and Planning, at the Delft University of Technology, with particular respect to the nature of Dutch land reform and spatial development. He comments that the very nature of sustainability and the long-term effects of production methods and emissions, may not yet be provable. Nevertheless, by failing to make judgements based upon ethical principles now, it may be too late to act later. However, by choosing the principle of sustainable development as a means of controlling the use of technical resources, it is possible to generate new problems in it's application to small-scale, short-term decisions. One way of overcoming this problem is to increase the level of education and research in the field of sustainable development and to incorporate it into the educational programmes. The Governing Board of the Delft Technical University have adopted a new strategic view termed, "towards a new commitment to society", which has served as a framework for the evaluation of all new curricula with respect to their educational and research role in sustainable development. This new strategy is discussed in terms of the curricula design for geodetic engineers and surveyors by addressing the roles played by that profession in urban and rural development. This is synthesised in the theme of sustainable development within spatial developments, which considers the role of the profession in communication throughout the planning process in both rural and urban regions, with particular respect to the implications of this role to education.

Weber takes a slightly different viewpoint to Rosman by suggesting that the logical key to sustainable development lies in spatial planning rather than through education and research. Spatial planning, it is argued, should be capable of acting in a central role as a mediator between the abstract concept of sustainable development and the very real problems involved in the physical processes of spatial development. In many respects, the paradigm of sustainability aims to optimise the interactions between nature, society and the economy in a way that will least affect nature. To achieve this objective, within any spatial development, a degree of cross-sectional orientation must be the starting point in resolving this problem. Since spatial planning comprehensively limits spatial functions and attempts to manage them, it is the key in placing the target of sustainability into practice. However, in attempting to achieve this goal, it must be accepted that the rules of sustainability can only be attained by the principle of proximity, in which the effects and contributions can be assessed within the dimension of an easily comprehensible area. Examples would include a defined region such as a municipality, a city district, a village, or a personal environment because it is essential that a good degree of identification is required to encourage local actions. With such local involvement in sustainability, Weber considers that spatial planning presents itself both as a suitable mediator and as an umbrella profession under which instruments and strategies, contributing towards the realisation of the goal of sustainability, might be ordered. This latter function might be exercised either as a way of presenting a variety of strategies for the mediation and accomplishment of sustainable action or in the form of prevention orientation, as an act of repairing the environment by decreasing, for example, the continuing the waste of natural resources.

Spatial planning also has a fundamental role in most countries in the debate on sustainable development. For example, in an urban context spatial planning will determine, very precisely, where and how buildings and technical infrastructures are located, powered, linked to the transport network, utilised, repaired and eventually disposed of. The same is also true in the case of rural development, land use and rural conservation practises. Weber comments that these factors may be summarised by relating sustainability to the quality of life in both a quantitative manner, in satisfying defined physical and spatial elements, and in a qualitative manner by generating a sense of comfortable living space. This relationship between sustainability and spatial planning may be considered to be symbiotic, in which the weaknesses of one can be compensated for by the strength of the other. Thus the whole reflects positively back on the individual parts.

Dixon-Gough *et al* (1998) commented that the European Commission's communication *"Europe 2000+: Co-operation for European Territorial Development"*, published in 1994, has made a significant contribution towards defining the role of spatial planning upon conservation and the sustainable management of rural areas. Spatial planning can act both as a constraining influence to protect habitats yet at the same time play a contributory role by identifying the benefits that might be accrued from the sustainable use of land. This is particularly so in socio-economic terms, whilst developing and improving the partnerships between local and regional authorities and those with economic and conservation interests. Furthermore, the same policy may operate at a strategic level to highlight some of the inter-relationships between different policies competing for the same natural resources. This policy will, inevitably, promote the concept of sustainable land use, ecology and conservation programme.

One of the principle challenges of sustainable development is that of reconciling the ambitions of various interest groups with the needs of the environment and of the individual (Elliott, 1994). Thus, the overall concept of sustainability is often very difficult to apply in practice and, unfortunately, there is a tendency for conflict to arise without one of the parties involved being aware potential conflict. Page (1997) gives an interesting example in which he contrasts the conflicting views of sustainability by a major supermarket chain in Great Britain. Apparently this chain is prepared to work with the Royal Society for the Protection of Birds (RSPB) to launch a scheme to save the skylark in many parts of rural Britain. He contrasts this with the free bus service offered by the same supermarket chain, which collects shoppers from outside the village shop to, "cart them several miles away to one of its superstores". He then comments that

> Surely a truly "sustainable " system is not just about skylarks, it is about the people who live and work in the countryside too, and about creating thriving, viable rural communities? One sure way to help a rural community to become *unviable* is by putting at risk the future of its village shops.

The main thrust of his argument was that a sustainable rural development should consider all aspects of the whole countryside – including the facilities and infrastructure necessary for rural people to sustain a good quality of life. It is not for nothing that the term 'rural poor' has often been used to describe the inhabitants of many rural areas in Britain, which exist without transport links, shops, post offices and many of the facilities and infrastructure that many urban populations take for granted. It is difficult to assess whether localised spatial planning rather than strategic spatial planning policies can satisfy the goal of sustainable development. In many instances, there is an under-riding element of cynicism in many proposals relating to sustainability, as commented upon by Rosman. This is particularly so in respect to agrarian reform (in the reform of the European Common Agricultural Policy (CAP)) and in the attempts by local government to demonstrate their policy of sustainability in respect to Local Agenda 21. This latter example demonstrates very commendable objectives that are, however, largely dependent upon funding to which the local authority does not have direct access.

The specific problems relating to rural and agrarian planning, identified above, are discussed in the chapters by Läpple and Seher. Both address issues that are related, respectively, to Germany and Austria yet, which also have considerable relevance to rural land reform, in general. In Germany the entire agrarian structure is undergoing change, which not only affects agriculture and forestry but also the entire rural infrastructure. The change has been brought about by two significant events; the unification of the two German states with their different systems of agriculture and types of society, and the reform of the (CAP). The adjustment from the socialist land law and collective farming in the former German Democratic Republic (GDR) to the market-orientated policies of the former West Germany, have required significant changes in land reform measures. When these have been combined with the European Union's (EU) rules on competition, the effects of the development (particularly in the rural areas of the former GDR) have resulted in many changes. In addition, many rural areas are experiencing

16

non-agricultural demands on land, such as the competing demands of urban development, industry and trade, transport, water, nature conservation and landscape management, as well as leisure and recreational facilities.

In order to preserve the economic strength of rural areas, Läpple defines the village as being the key element. In Germany, the village is being actively promoted as an alternative settlement area to living and working in urban conurbations. Providing the population density is maintained at a sufficient level, the basic infrastructural demands can be retained and even developed. Since these facilities include schools and village community centres as well as shops and banks, this policy is something of a contrast to Page's (*op cit*) experience in rural England. Läpple considers that it is important to develop the village as an alternative settlement since they cannot wholly be sustained by agrarian economy. If, in order to pursue a policy of rural sustainability, services in environmental and nature conservation together with landscape management are to be maintained, alternative forms of remuneration must be sought. The key to this is the sustainable village community and a programme based upon a policy of integrated rural planning within regional economic policies.

As well as the demographic changes currently taking place in both rural and urban areas, there have recently been a number of structural adjustment programmes, world trade agreements, international credit arrangements, which have linked the agrarian economy to the world economy in a way never previously seen. It is also becoming increasingly evident that the majority of the world's rural populations cannot subsist on farm activities alone. Food processing, transportation, urban and other non-farm activities are providing an increasing proportion of the agricultural family income. In Britain, livestock farmers have been assailed by the BSE crisis, the strong pound and competition from products from other parts of Europe and the world. In many cases, livestock farming can no longer be considered to be an economically sustainable and viable activity. Livestock farmers are currently leaving the land at the rate of 20 a week and many of those that remain need another form of employment in order to survive.

The present crisis in agriculture is being felt across Europe. It is largely the result of a surplus in agricultural production which has, in turn, led to a general decrease in the number of farms. Since the rural landscape has essentially been shaped by the agrarian economy, the crisis in agriculture and the resulting decrease in agricultural income will result in the land being affected by the need either to increase its productivity or to be utilised for purposes unsuited to the location. Seher comments that the agricultural and natural habitats, previously been considered as simple by-products of the agrarian economy are likely to become endangered. He cites the example of marginal farmland unsuited to intensive cultivation, which often has a dual nature with regards to the preservation of recreational areas and the protection of habitats from natural disasters. Any retreat of agriculture from, for example alpine pastures, might have far-reaching consequences upon the rural economy and upon the many non-productive agricultural functions. Seher proposes that any future agrarian policy must not only consider the stabilisation of farm income, but also the rehabilitation and improvement of the natural environment in a sustainable manner.

Against recently defined agro-political objectives, the Department of Land Consolidation in Lower Austria has developed an Ecological Scores for Agriculture Model (ESAM) as a means of targetting rural and agrarian aid. The ESAM is primarily directed towards the production-independent direct payments, the ecologisation and extensification of soil cultivation and to honour existing ecological outputs. This is achieved by quantifying the ecological output of agriculture within a pre-defined set of scores. Seher describes the concept and operation of the ESAM with particular respect to its application as a suitable instrument for supporting sustainable cultural landscape development.

The twentieth century has witnessed an enormous acceleration of urbanisation. The proportion of urban population to total population stood at about 16% during the nineteenth century. By the end of the twentieth century, it is expected to reach almost 48%, 50% by 2005 and 61% by 2025 (UN, 1995). This population increase in urban areas, within both developed and developing countries, has created a series of environmental, social, cultural and economic problems. Urban growth rates have frequently exceeded the capacity of the infrastructural and institutional mechanisms to support them.

The process of urbanisation generates a need for modifications at all aspects of urban and rural life, even to ensure the basic provision and distribution of food and water. Food for urban dwellers becomes more expensive as transportation and distribution costs increase and because a far greater proportion of processing costs are involved. One of the many interfaces between the rural and urban environment is that of tourism and this is investigated in the following two chapters by Zwirner and Willaims who examine the concept of sustainable tourism. Zwirner argues that sustainable development is a concept that draws upon two frequently opposed intellectual traditions; one concerned with the limits which nature presents to human beings, the other with the potential for human material development locked up in nature. In exploring the concept of sustainable tourism, Zwirner introduces the relationship between sustainability and industrial redevelopment within a tourist region. The site in question is a large steel mill situated on the edge of Lake Iseo, in northern Italy, although the scenario might be recognised across Europe and throughout the world. The closure of the steel mill was largely the result of unsustainability through difficulties brought about by the combination of increasingly heavy goods vehicles and unsuitable roads. Zwirner proposes a number of developments that would satisfy the criteria of sustainable tourism, yet would also be acceptable to both the commune and the local people and which would also blend into the physical fabric of an essentially rural region. In turn, the developments must satisfy the demands of a transient, yet essentially urban population of tourists.

A similar case is presented by Williams who concentrates upon the environmental changes connected with the construction and use of marinas within the fragile coastal zone. With any such development, the prime location would be in a 'rural' rather than urban area, although many urban marinas exist, examples being those in Barcelona and Swansea in what were formerly ports and docks. Irrespective of the location of the marinas, Williams suggests that the needs of the local community should be taken into consideration when designing the marina a full environmental impact assessment be conducted to reveal possible adverse factors relating to its siting and use.

Communities and human resources are one of the most important factors in determining the productivity of a population and its behaviour. Basic investments in education, health and social infrastructure will lead to changes in behaviour, improved work productivity and will generate development. They are the primary means of ensuring environmental sustainablility and human and economic development.

Although the concept of sustainability is frequently associated with rural areas, UNEP (1997) makes reference to the concept of the sustainable city. This is defined as having a lasting supply of natural resources on which its development depends and a lasting security from environmental hazards, which might threaten development achievements. A sustainable programme of urban development supports conventional planning and

19

management objectives by making efficient economic use of development resources, and by providing a level of social equity in which the benefits and costs of the development are equitably distributed. Furthermore, it must clarify environmental issues, agree on joint strategies and co-ordinate action plans, implement technical support and capital investment programmes, and institutionalise a continuing environmental planning and management routine.

The remaining chapters in the book provide examples of these issues relating to land reform and sustainability within the urban environment. de Wolff considers the issues of sustainable urban development in the Netherlands, by examining the measures of land reform adopted by central government as part of the process of designating new urban areas. The process of the 'new urbanisation concept' has been developed in the Netherlands on the basis of two assumptions; a reduction in the need to travel, and rural areas need to be protected from further development. As part of the new urbanisation concept, 26 urban regions have been identified by central government and outside those regions, the development of industrial and residential sites will only take place if required by the local population. The new policy identifies two types of new urban region; those located within existing urban areas (inner-city or 'brown-field' sites), and those directly connected to existing urban areas. Integral within this policy is that of the 'durable building' concept within a high-density layout designed to minimise the impact of the new urban regions upon the urban/rural fringe. However, de Wolff acknowledges that this form of sustainable urban development is at odds with the single family or, more specifically, the 'house with a garden' that is still the most favoured form of dwelling amongst the majority of the population of the Netherlands. High-density urban areas are, however, not only desirable for the protection of open spaces, but they are pre-requisite for the development of good and frequent public transport facilities.

The changes in policies related to the development of new urban regions in the Netherlands is further complicated by the decline in the traditionally dominant role of local government in the land market. This, in turn, makes the realisation of the concept of urban sustainable development even more complicated. The final test, however, must lie with the consumers who must be convinced that a sustainable location is preferable to a traditional one.

20

In contrast to the concepts associated with new sustainable development being described by de Wolff, Dixon-Gough examines the concept of the 'railway town' and, by taking the example of Swindon (located in central, southern England), outlines the periods of sustainable growth that have taken place. The railway town may be an example of both unsustainable and sustainable urban development. Some towns, such as Melton Constable in East Anglia and Caerphilly in South Wales, have declined in importance with the corresponding decline in the importance (or removal) of the railway. Others, such as Swindon, have continued to develop, expand and increase in importance after the prime function of their existence has declined. In the case of Swindon, the 'New Town' was developed to serve the needs and to provide services for the workers, and their families, of the Great Western Railway's works. At the end of the nineteenth century, Swindon was essentially a one-industry town. Diversification commenced during the early twentieth century, encouraged initially by the railway company, latterly by the Borough Council and most recently through the aid of grants from central government. The railway works closed in 1985, but the town continues to experience sustained growth with the emphasis now being in the service sector rather than in heavy engineering.

History has shown that there has been a close relationship between urban areas and violence. To many, violence and trouble-making within urban areas are post-war phenomena but urban riots and insurrections existed and are documented in ancient Greece and Rome, and were commonplace within medieval cities. Doughty makes an examination of inner-city violence, which is often the result of social unrest and largely manifests itself in high-density residential areas that are frequently populated by ethnic minorities. Doughty recommends that the sustainable development of inner city areas requires the combined efforts of environmentalists, planners, politicians and must, above all, be sympathetic to the demands of the residents who will, or will not, benefit from the development. de Wolff also alludes to this factor.

In the chapters by Dixon-Gough and Doughty, the concept of sustainability within urban areas was not linked to land reform *per se*. Home, however, describes the process of sustainable development within the Thames Gateway, which links land reform (through the processes of restructuring and remaking the urban landscape) very closely to sustainability through the London Dockland regeneration project and

21

the subsequent concept of the Thames Gateway. This region has a long history of environmental degredation, being situated down-river and down-wind of the City of London. In an era relatively free from regulatory controls, this area became the concentrated location of many noxious industries and most of London's utilities, including the largest gas works in Europe, the largest coal-fired power station in Europe and the end location of the first comprehensive drainage and sewerage scheme for any major city. Most of these activities have now ceased, leaving behind a legacy of derelict and contaminated land. Home, outlines the 'new environmental standard' sought by the Thames Gateway planning framework, in which the region may be utilised to provide an alternative to the development pressures being experienced elsewhere in the environmentally superior areas of the south-east of England. Inclusive within this policy will be the environmental reclamation of the riverside land, contaminated and spoilt by past industrialisation.

In keeping with the concept of abandoned land, Bullard defines the broad categories, whilst identifying the conflicts that frequently emerge when attempting to develop a strategy involving their sustainable development through reclamation.

The dynamics of land reform and sustainable development are increasing as we come to the end of the twentieth century. In part, these dynamics are fuelled by the need for social justice and compensation and for what has been inflicted upon the land and its population in the past. The needs of market forces also play an important part in these dynamics as the forces of globalisation inexorably affect our lives. However, until the spheres of social justice and market forces begin to overlap, the quest for equitable land reform and true sustainable development, in both rural and urban areas, will continue.

References

Boyle, N. & Wright, A. 1992.   Infrastructure notes, *Urban No. OU-5*, Transportation, Water and Urban Development Department of the World Bank, Washington D.C., USA.

Brimicombe, A.J. & Yeung, D. 1995.   An object orientated approach to spatially inexact socio-cultural data. *Proceedings of the 4th International Conference on Computers in Urban Planning and Urban Management*, Melbourne, **2,**

519-530.

Brooks, K.M. & Lerman, Z. 1994. *Land Reform and Farm Restructuring in Russia*. World Bank, Washington D.C., USA.

Cullingworth, J.B. & Nadin, V. 1994. *Town and Country Planning in Britain (11th Ed.)*. Routledge.

Czaki, C. 1994. Where is agriculture heading in Central and Eastern Europe? Presidential address to the *XXII International Congress of Agricultural Economists*. Harare, Zimbabwe.

Dale, P.F. 1997. Land management: global problems, local issues. Keynote Address at the *International Conference of Land Management* held at the Royal Institution of Chartered Surveyors, London. January, 1997.

Dixon-Gough, R.W., Mansberger, R., Bullard, R. & Seher, W. 1998. Global, regional and local policies for land conservation and planning. Presented at the *27th International Symposium of the European Faculty of Land Use and Development*, Zurich, May 1998.

Elliott, J.A. 1994. *An Introduction to Sustainable Development*, Routledge.

Gordillo, G. & Riddell, J. 1996. *Habitat II and the World Food Summit: the challenges ahead.* (07/14/97)

Jankowski, P. & Stasik, M. 1997. Design considerations for space and time distributed collaborative spatial decision making, *Journal of Geographical Information and Decision Analysis*, 1(1), 1-8.

Litven, D. 1998. Development and the environment, *The Economist*, **21st March 1998**, 3-16.

Page, R. 1997. Country diary, *Weekend Telegraph,* 12th July 1997, 14.

Thamesdown Borough Council, 1996. *The Seeds of Change. Thamesdown Borough Council's Contribution to Local Agenda 21*. Civic Offices, Swindon.

Thiesenhusen, W.C. 1994. *Landed Property in Capitalist and Socialist Countries:Nature of the Transition in the Russian Case*. Land Tenure Center, University of Wisconsin, Madison, USA.

United Nations, 1995. *Compendium of Human Settlement Statistics*, UN, New York.

UNEP, 1997. *Sustainable Cities Programme.* (06/12/97).

WCED, 1987. *Our Common Future*, Oxford University Press, Oxford.

2 The naturalists' dilemma and some ecological consequences for development

H. LENK

It is questionable whether economic thinking and economic models are adequately suited to measure ecological damages. Should they be applicable at all, they would have to be taken for granted that the monetary value of everything can be unambiguously determined to the nearest penny. How, for example, might ecological or irreversible damages, or impairments due to long term effects that cannot (or would not yet) be recognised today, or even the intrinsic value of nature be measured? In order to determine forest damages, what should and would really be measured: the individual owner's loss of profit or the lost benefit to a community that is, in principle, open? Moreover, how can we decide on costs and benefits of future generations or even anticipate their criteria of evaluation? (Besides inflation how can we discount future benefits as discussed? Is it morally legitimate to discount anyway?) Damages to ground, air, water and health, which are caused by external negative effects and other social and ecological consequential damages are not actually subject to easy measurement in monetary dimensions. Indeed, this seems to be the one prominent reason why they are systematically neglected in the usual GNP. How should the extinction of species, quality of spare time, public goods etc. be measured in monetary value?

Ecological damages can usually not simply be attributed to a single (individual) producer or responsible person. For example, in 1989 a hearing of the Economic Commission, in the Lower House of the German Federal Republic (Bundestag), assessed the external costs of environmental damages. These concerned only the field of motor traffic and were assessed by the director of the Federal Agency for Environmental Protection (Umweltbundesamt), Lutz Wicker, at the level of approximately 50,000 million German Marks. This, however, also does not take into consideration the attribution of individual damage, from somebody who is, or many who are personally responsible although car traffic is usually considered to be essentially an individual affair in the first place.

(Incidentally, the total assessment of man-made damages to the ecology for the very same, relatively small, country was assessed to cost more than 103,500 million German Marks).

The problems of attributing responsibility are to be found, in particular, today in highly developed industrial societies shaped by technology and advanced economies. Personal action seems to disappear behind collective and group actions. The individualistic concepts of ethics and philosophy, technology and economy do not suffice to tackle these problems. They are evidently not adequate, since they usually focus almost exclusively on individual actions and not on interactional, collective and corporate forms of actions, or structural and systematic contexts. Thus far, ethical approaches have indeed been far too oriented toward individual persons, have not paid enough attention to social aspects, and are thus not adequately adjusted to socio-ethics and socio-philosophy. This paper examines some methodological questions of economic assessment with respect to the detrimental ecological encroachments on the environment, and paradoxical developments with respect to built-in dilemmas of escalating social problems, like the so-called "Tragedy of the Commons", the "Free-Rider Problem", the "Naturalists'" or "Enjoyers' Dilemma", are analysed.

It is true that so far the problems of complex constellations of causes and the problems of responsibility have only been discussed only in a very generalised way in philosophical literature. However, the jurisprudence is considering them in a much more detailed manner and has indeed come up with some very interesting approaches for solutions, which are of interest for philosophical reasons as well. However, as a resume - and in a modifying and restricting sense - it must be mentioned that the actual convincing principle of causation faces difficult problems, if taken as the only precondition for the attribution of responsibility. These difficulties result primarily from the fact that there are various forms of collective action. Moreover, there is a far-reaching impossibility to individualise causal factors within synergetic and cumulative processes due largely to the formation of groups and to the effects of adverse combinations of many intruding factors. To a large extent, legal regulations (de *lege lata*) fail as ways for obtaining adequate precautions or even solutions when faced with ecological damages and damages that occur far from the sources of emission. Often a need for legal regulations is acknowledged. Yet for strategic environmental planning and policy-making it is so far only insufficiently considered, nor has it until now been rendered controllable or sanctionable. In the literature one finds discussion and propositions about joint (and several) liabilities. These include: a mutual right to compensation through the formation of sensitive areas and feedback loops of danger, products and strict liability; the turn-about of the burden of proof; high probabilities of causation; compensation through special public funds; (structural) incentives for the internalisation of external effects; environmental emission bonuses to be bought by corporations, etc. The main difficulties for legal solutions are, among others, the extant and non-liability connected with permitted and subliminal (damaging) individual actions, and the definition of limiting and threshold values.

It is, in particular, the responsibility for usage of land and nature, of their respective systems, ecosystems and species, the responsibility for maintaining ecological balances, the co-responsibility for survival conditions of future generations, for maintaining, regenerating and moderately (as well as reasonably) utilising resources and the responsibility for side-effects and remote consequences of actions which are dramatically gaining increasing moral and political relevance.

However, joint responsibility (co-responsibility) is not a cake that can be shared. Mellema (1985a) proposes the actualisation of obligations by linking them with threshold values; he thus wants to turn against the effects of ethical dilution through, as Jaspers put it, "Mitläufer" (fellow-traveller or easyrider), the excuses of being "forced by orders" and the growing numbers of members in groups with joint (collective) responsibility. In general, moral responsibility cannot be shared like a cake and diluted like a cocktail, even when the impression might occur within political contexts, as e.g. in elections, parliaments and committees, that the growing numbers (of members) would reduce or even dilute the very responsibility of the participating individual.

The pie-model ultimately proves to be too simplistic and unfeasible. Real cases are usually not that clear-cut in the first place. In addition, explicit knowledge of the situation, readiness to co-operate or refuse, "we-group" feelings and "team effects" must also be considered.

In what follows, we would like to deal with a type of distributability problems, viz. the responsibility in collective acting. In this regard we have consequences, impacts and side effects of non-corporative collective actions (e.g., the strategic market oriented ones). In economics and social science scholars speak of the externalities problem, side effects, social costs, social traps, the "Prisoners' Dilemma", and the public goods problem.

Externalities are "neighbourhood or third-party effects(s) of an exchange" (Buchanan, 1985, p. 124) or of other economic activities (e.g. production or consumption). The externalities problem is present in the regulating, attributing, accounting for and abolishing of external effects. Generally speaking, we have market-oriented economical, governmental or state-oriented and co-operative solutions for the problem. If externalities ensue, private property rights are insufficiently or not determined at all, or sometimes not even asserted or claimed at all. Such property rights guarantee the exclusive usage or disposal· of goods and rights; connected with property rights are, in some cases, the costs implied in utilising them, i.e. the so-called internalisation of external effects. Voluntary internalisation of externalities would, however, presuppose that the costs of internalisation are much less than this amounts to a further problem.

Public goods are:

> any desired state of affairs that satisfies these conditions: (i) efforts of all or some members of a group are required to achieve the good; (ii) each member of the group regards his contribution as involving a cost; (iii) if the good is achieved, it will be produced in such a way as to be available to all members of the group, including noncontributors (jointness of supply); and (iv) if the good is produced it will be impossible or unfeasible to exclude noncontributors from partaking of it. (If, in addition, one individual's consumption of the good does not decrease the amount of the good available to others at all, the good is a *pure* public good). (Buchanan, 1985, p. 125).

Whereas responsibilities are clearly determined (usually by law) for exchange on the market, where the exchange of individual goods and services is concerned this does not, at least ideally (necessarily) apply to the usage of collective goods (because there is no symmetry of exploitability and exchange). Morality and law are themselves examples of public goods; there are social traps with respect to them too. The problem could be illustrated firstly by using the structure of the so-called "Tragedy of the

27

Commons" (Hardin, 1968). This constellation can be understood as a prototype of a social trap. (Does not the extant threat of overpopulation figure as a major or even a social trap for developing countries in the first place, but also for mankind in general?) The central question will turn out to be, "Who would bear the responsibility for the result of an action result and for the respective consequences, which nobody had wanted or intended beforehand?"

According to Hardin (*op cit.*), every owner of stock in the Sahel zone has an individual and perfectly legitimate interest in utilising and exploiting the common grassland, the so-called "commons", which is indeed a collective good. This individual interest is characterised by striving to have as many stock as possible, because the greater one's own stock, the higher is one's social status. All the owners and society in general, however, have a common interest (a real commonality) namely to avoid overgrazing of the commons. This constellation of individual and common interests would lead to the following dilemma:

> Because nobody has sufficient individual interest to avoid extensive exploitation of the commons for his own good, everybody will utilise it as extensively as possible, thus overgrazing of the commons would be the necessary result and consequently, in the last analysis, the very satisfaction of the individual interest would be barred or ruined, respectively.

Hardin (*op cit.*) considers it necessary to have social, i.e. non-individual mechanisms of control, in order to avoid such a dilemma. Socially enforced co-operation would, for example, be such a controlling mechanism. He emphasised that such "tragedies of the commons" would undermine or at least relatives the well-known traditional theorem of "the invisible hand". ("The invisible hand", in terms of the market mechanism would, according to the opinion of classical and neo-classical economists, results in a constellation that the consequences (profit or loss, respectively) would be attributed to the responsible actor and that an optimum overall result, in terms of an optimal equilibrium and general wealth, i.e. a Pareto optimum, would occur.) According to Hardin (*op cit.*) the rational maximising of each individual interest need not, via dynamic market processes, lead to an optimum result and wealth for all. On the contrary, it may lead to depletion, erosion and pollution of the common land. A similar problem, with respect to arable land use, also leads to depletion, erosion, even devastation of arable land in large parts of Africa: the few remaining trees and shrubs are necessarily used and/or consumed to satisfy pressing survival interests of individual families. This consumption leads to further expansion of the desert and to an additional deterioration of sustenance and survival conditions of the whole population etc. (With respect to stock and the above-mentioned traditional conflict between the individual owners' interests and social needs, even the boring of additional wells might even aggravate or escalate the conflict constellation and accelerate the ecological problems. This might be a well-known unintended side-effect of political and economical development programmes.)

A similar effect is the clearing and conversion of tropical rain forests to arable land on basically poor soil, which might lead to local and regional erosion and eventually to the depletion of the ecosystem. This, in turn, might eventually lead to a continental or even global change of the climate (cf. the global carbon dioxide problem and the impending green house effect of overheating the atmosphere).

According to Hardin (*op cit.*) the problem of environmental pollution turns out to be of an analogical or equivalent structure. The commons, a public good in this case, however, is not diminishing or decreasing in size, but instead a negative quality is added, namely through the depositing of refuse of many kinds. Again, it is profitable, i.e., cheaper for the individual actor to do away with garbage on public soil, e.g., to deposit chemical refuse in the Rhine. As a consequence of these public measures, external social costs would result. Negative external effects would amount to a burden for the general society; they can only be avoided or redirected if the taxpayer or everybody pays in money or is suffering in terms of health disadvantages, deterioration of quality of life or aesthetic qualities of ecosystems and the landscape. Externalities would result from the actions of producers and consumers whenever these agree on actions which would be disadvantageous for the environment (think of the example of the one-way bottles). There is, therefore, a growing and pressing responsibility of the consumers, a co-responsibility with respect to the protection of the environment. On different levels of a scaling phenomenon, *all* members of a society would bear a certain responsibility.

Generally speaking the same structure is to be found with many problems of social constellations which can be dubbed, "social trap constellations". It would be profitable for individuals to infringe social rules and norms as long as (almost) all other members abide by them. A similar structure is to be found in the so-called, "free-rider" and "assurance" problems with respect to providing and maintaining collective and public goods. Both cases lead to social traps. The dilemma of environmental protection on a voluntary basis is an intriguing example of this constellation. The free-rider problem is:

> a barrier to successful collective action or to the production of a public good that arises because all or some individuals attempt to take a free ride on the contribution of others. Non-contributors reason as follows: either enough others will contribute to achieve the good or they will not, regardless of whether I contribute or not; but if the good is achieved, I will benefit from it even if I don't contribute. Consequently, since contributing is a cost, I should not contribute. (Buchanan, op cit., p. 124).

The assurance problem is:

> a barrier to successful collective action or to the production of a public good that arises when all or some individuals decide not to contribute to the good in question because they lack adequate assurance that enough others will contribute. (ibid.).

The provision and maintenance of a collective good is, according to Olson (1968), primarily dependent on the magnitude of group membership; the greater a group of participating individuals, the less the chance turns out to be for providing and maintaining such a good and the greater is the necessity of compulsion, sanctions etc., with respect to usage and distribution of collective goods. Whereas community norms or a morale would still seem satisfactory for reaching a common goal in small groups, this does not apply to large groups. (James Buchanan called this phenomenon "the large number dilemma" (Vanberg, 1982, p. 137).)

The structural problems of social and individual actions, of public goods, and of the commons and social order can easily be illustrated by using the well-known game and theoretical model of the so-called Prisoners' Dilemma (PD). In the classical situation of the Prisoners' Dilemma, two prisoners (A, B) are indicted for armed robbery. Both are offered to be the chief witness and to be released free without penalty. Individually, they do not know about the other's offer and there is no communication between the two of them.   Both of them can only be convicted because of illegal possession of weapons.   Therefore, if both remain silent, they can only expect a rather minor punishment (of, say, one year in prison), but a much higher punishment (ten years), if one is convicted (punishment for the chief witness would be zero).   Confessing, therefore, seems to be preferable as the dominant strategy.   If one of the two confesses, it is also profitable for the other one to confess, because then he would receive eight instead of ten years imprisonment.   The amount of punishment is, therefore, not only dependent on one's own strategy, but also on that of the co-prisoner.   Now the dilemma of the social trap consists of the fact that it turns out to be irrational for A as well as B, in their own interest, to confess (dominant strategy).   But if both of them act rationally, i.e. both would confess, they would incur a higher punishment (eight years each in prison) than if both were keep silent (only one year each), i.e. if they would act co-operatively.   Individual rationality, therefore, leads to collective irrationality and self-damage, to a pareto-inferior state.

> Pareto Optimality Principle (most inclusive form): A state, s1, of a system is Pareto Optimal if, and only if, there is no feasible alternative state of that system, s2, such that at least one individual is better off in s2 than in s1 and no one is worse off in s2 than in s1. (Buchanan, op cit., p. 125).

Sen (1987, p. 86) characterises the strategies as:

1.   co-operative strategy: it is better for the respective goals of all of us;
2.   non co-operative or defective strategy: it is better respectively for each of us, given what others do.

We might easily conceive a positive variant of the PD which could be referred to as the "Naturalists' Dilemma" or, in more general term, "the Enjoyers' Dilemma" (ED) with respect to scarce resources.   Imagine the only lake in a nature-reserve (e.g., in a US National Forest) which is to be enjoyed and partially utilised by anglers (A) and water-skiers (S) at the same time. For the sake of argument, the lake should not be large enough for both to enjoy their sport at, or on, the lake without interfering with the other.   If both the anglers and the water-skiers used the lake in an unrestricted manner, they would not to be able to enjoy their sport on the lake.   The water-skier would mingle with the anglers' lines frightening away the fish from the anglers and, at the same time, introducing the possibility of damage to themselves and/or the boats as a result of entanglement with the anglers' lines.

A,B	*s*	*c*
s	*1,1*	*10,0*
c	*0,10*	*8,8*

$$\left. \begin{array}{l} \textit{for } A\!:B\,s \rightarrow A\,c\!:0 \leq 1 \\ \textit{for } A\!:B\,c \rightarrow B\,c\!:8 \leq 10 \end{array} \right\} Ac \\ \left. \begin{array}{l} \textit{for } B\!:A\,s \rightarrow B\,c\!:0 \leq 1 \\ \textit{for } B\!:A\,c \rightarrow B\,c\!:8 \leq 10 \end{array} \right\} Bc \right\} 8,8\, \textit{instead of } 1,1\}$$

s: silence, c: confession as action strategies
ammounts: years of imprisonment (first for A, second for B)

Figure 2.1 The prisoners' dilemma

Thus, an agreement would need to be reached with one another to enable both to enjoy a nature reserve. They would have to arrange for restrictions by, for example, segmenting space or time. Part of the lake could be allotted to the water-skiers and the remainder to the anglers or they might permit water-skiing and angling on alternate days. Other possibilities of restrictions are conceivable. However, any restriction and segmentation, whatsoever, would decrease the full-scale enjoyment of both parties. Therefore, the dilemma which arises does not develop from the bargaining of negative sanctions as in the classical PD, but as a dilemma of the full-scale enjoyment with respect to a scarce nature reserve. In this variant, it is not the sanctions of the object of the potential agreement, but the possibility and degree of positively enjoying the natural source or resource that are at stake. It is largely the same concept that Hardin (*op cit.*) had in mind with the overgrazing of the Sahel-zone mentioned above. The most important difference from the PD is that in ED situations, tiered possibilities or levels of opportunities of utilisation do occur. This is in contradistinction to the yes-or-no strategies involved in the PD model which admits to variations with regard to degrees or intensities of utilisation or even by partially dispensing with them. Here, the pay-offs may, within limits, be determined at will or chosen by steps.

S,A	a	¬ a
s	0,0	4,0
¬ s	0,4	0,0

for A: S s $\varnothing$ A indifferent (0=0)   *(1)*
      S¬s $\varnothing$ A a   *(4>0)*

for S: A a $\varnothing$ S indifferent (0=0)   *(1)*
      A¬a $\varnothing$ S s   *(4>0)*

Figure 2.2 The enjoyers' dilemma

It might be as difficult to reach an agreement in this sort of Enjoyers' Dilemma, as in the classical Prisoners' Dilemma. It is, however, not just a change in signs in the respective utility and evaluation functions, but the different points of equilibria that might occur. For example, when dealing with the above situation of the water-skiers and the anglers, it might be assumed that just one, instead of several water-skiers and just one, instead of many anglers would like to use the lake. By presupposing the correct size of the lake, there is no loss or restriction of generality with this assumption of just one representative in each set. The values of the respective net output matrix are assigned so that the following fictitious matrix would result (a: angling, -a: not angling, s: water-skiing, -s: not water-skiing). If the maximin criterion is interpreted as > or <, respectively, there can be no acceptable strategy for a general solution because of (1). If

(1) is disregarded, however, (s, a) turns out to be an equilibrium point (with 0,0). Should the skier and the angler use the lake only in special conditions (c) (conditioned corporation), e.g., interchangeably every second day, the following matrix would result:

S,A	c	a
c	2,2	0,0
s	0,0	0,0

for A: S c ⋈ A c   (2>0)   (1)
    S s ⋈ A indifferent (0=0)

for S: A c ⋈ S c   (2>0)   (1)
    A a ⋈ S indifferent (0=0)

  If the maximin criterion is interpreted in a strict sense again, it would again not arrive at the general strategy of a solution - because of (2). This would, however, be completely fictitious. If (2) is left out of consideration, (c,c) would be an equilibrium point (with 2,2).
  Therefore, if S and/or A should use the lake exclusively, it would follow that:

$$U_s = O = U_A \qquad \text{(U: utility)}$$

a meaningful strategy for S or A would, however only be s or a respectively, because only in this way could U > O. Conditioned co-operation is, therefore, the logical strategy, because only in that way can "pleasure" (utility) originate at all. This is the reason why any discussion concerning the Enjoyers' Dilemma involves a small structural difference in comparison to the classical Prisoners' Dilemma, because attention has to be paid to some additional pre-suppositions with respect to the content (defying or avoiding zero utilities for both parties). Levelled gratification's can result, if both partners dispense with parts of their enjoyment. In this case, even a positive balance of enjoyment may be maintained (e.g. using the lake every second day). Co-operative behaviour, with respect to abiding by agreements, treaties or contracts, can turn out to be useful for both Parties.

In addition, a strict maximin criterion does not avail itself of a unique decision; additional criteria are required. (Moreover classical PD-situations are static with respect to the output matrix, whereas a dynamic output with respect to the degree of and intensity of utilisation was given above). Such additional criteria (e.g., conditioned altruism, enjoying the co-operation and stabilised agreements in an otherwise ever-changing system of utility conditions, etc.) would certainly allow the model to be refined and decision criteria to match reality more closely). It has to be considered, however, that the so-called "large-number dilemma" cannot be solved and that in large groups, lower transactions costs (cost of information, bargaining and checks and control) have to be taken into consideration.

Generally speaking, the positive variant of the Enjoyers' Dilemma seems to be of considerable interest besides the classical and static PD restricted to a bargaining of negative sanctions. The Enjoyers' or Naturalists' Dilemma seems to apply not only to the use of common land or nature reserves, but also to privately owned and exploited land if it is embedded in an endangered ecological environment. The groundwater level as well as clean air, drought or polluted air, erosion and depletion would not stop at a conventional borderline, but would affect the whole local, regional or even continental ecology. Thus, the Enjoyers' Dilemma seems to be a fairly general model with regard to the use of land and environment.

A detailed analysis of the PD structure shows that strategic actions of competing, self-interested, rational actors leads to a result which turns out to be an unintended social consequence placing all participants on a worse level than a co-operative strategy of abiding by social rules would have obtained. PD constellations cannot be solved on a pure individualistic level.

The above-mentioned dilemmas are also examples of rationality traps: the individually rational action strategy leads to collective social irrationality, undermining the first one. Under certain conditions, individual rationality can be self-destructive.

The distributability problem of responsibility consists in the fact that side-effects cannot be attributed to a single originator and that they usually were/are or even could not be foreseen or predicted. Two partial problems can be identified here: firstly, the question of participatory responsibility with respect to cumulative and synergistic harmful effects and, secondly, the question of how to deal responsibly with unforeseen or even unpredictable facts or side effects. The first problem can be termed the problem of distributing responsibility under strategic conditions. For example, is the legal principle of attributing "causality" and responsibility valid in Japan since the case of the Manama disease, which according to the statistically assessed contribution to the common harm by relevant polluters in the vicinity is ascertained as the pertaining causality by law, indeed satisfactory?

The burden of proof here lies, so to speak, on the side of the potential originator, the polluter, who has to prove the harmlessness of his emissions. This reversal of the burden of proof seems to be, at least, a controllable and operational measure to allow for attributability wherever environmental damages are in question. In this type of problem, the effects are usually the result of a combination of land, water and air use or misuse. They can at least be tendentially forestalled or diminished in a controllable way by assigning sanctions. In that respect, the Japanese legal principle of attributing causality might foster environmental protection. But there are methodological and legal, as well as moral problems connected with such a regulation. Firstly, adjacency and the guessing of causality can never be a proof of a causal origin. In addition, a simpler part of the problem is how to attribute distributively the responsibility in the cases of synergistic and cumulative damages, particularly those with below-threshold-contributions of individual actors. A further problem is how to distinguish between a descriptive assessment of causal origination and the normative attribution of responsibility, between causal responsibility and liability, Hart (1968). How could one possibly distinguish between the causal impact, the descriptive responsibility (Ladd, 1975), i.e., the descriptive attribution of responsibility, and the respective normative attribution of responsibility for contributions - the amount of which is individually ineffective, below the threshold of harmfulness? And how is one to distribute this kind of responsibility in general? Would it not be meaningful to postulate a normative collective responsibility of all pertinent corporations within the respective region in the sense of a joint liability? This would, however, mean a liability of all relevant corporations for the total damages. The impaired parties could sue for damages, claim in court for compensation and/or indemnification from any presumably participating corporation. Does this make sense, if connected with an overall generalisation? This regulation, however, would have the advantage of dispensing with the proof of damage in respect of each singular damaging or aggrieving party - as for example, a respective norm in

German Civil law would prescribe (cf. BGB §§ 830 I, 840 I, 421, 426). This kind of regulation would, in some way independent of individual case argumentation, interpret all non-collective agents as quasi one corporative agent being liable in total. The internal distribution and compensation within this quasi-group of corporate agents would then be a problem of mutual bargaining of all aggrieving parties.

Social traps can arise within the contexts of joint (and severable) liabilities, especially with respect to liability funds. Such traps could and should be prevented through specific regulations, e.g. regulations of premia or (positive as well as negative) incentives in correspondence with the potential hazards and damages, rights to compensation against single damagers' etc.

Notwithstanding these arguments, another kind of total liability with respect to product safety and hazards in terms of environmental damages of public goods should be established. It should be noted that there is a European Community agreement (1985) with regard to product liability laws. Causal originators of damages would then/now be liable in the sense of a strict liability in tort, whether or not they are really guilty in terms of intent or only negligent. Causal origination would already ascertain descriptive causal action responsibility and with respect to the damage of a good to be protected also normative responsibility for the respective action and its consequences. This form of liability would hopefully be deterrent enough to prevent infringements. If, however, damages would nevertheless occur it would at least not be necessary to prove fault or guiltiness as a presupposition of any claim for compensation. These arguments are equally relevant for the second partial problem with regard to non-corporate action.

The proposition for putting the proof upon the defendant, for a change of the burden of proof, which is not new indeed, should not be understood as a strict principle, but as a guideline for cases that must be defined more clearly. Legally speaking, if the potential damagers are known - and here they have at first to be considered as being faultless - i.e., if there exists a statistically significant connection between the damages and the releasing/causing occurrence (e.g., emissions) or if the connection cannot be explained better, which would be the injured/harmed person's burden of proof (cf. Weidner 1985, pp. 97), then the potential damager has to bear the burden of proof (and this is not yet to be considered as a sufficient reason for a constitution of guilt in the first place!). The potential damager would now have to prove that he or she was not the cause of the damages, either alone or partially. The following justifications might be mentioned, for such a regulation:

1. whoever benefits from an activity **has to bear** the risk of damage ("principle of cost and benefit of product liability" according to Diederichsen & Scholz (1984, p. 38));
2. whoever takes a risk or puts a burden on someone else in terms of a risk or an increased risk, is *acting* and should at least bear the burden of proof;
3. if "the impossibility for clearing up the causes originates from the sphere of the emitting sources or agencies the burden of proof remains with them ("theory of spheres" (Diederichsen/Scholz, ibid.)).

The legal problem of the so-called "proof until the doorstep" turns out to be similar: investigations should be performed only up to the doorstep of the respective plant or firm (of the defendants), since only the defendants themselves know and are to be held responsible for what is happening within their walls (Weidner, 1985, p. 99). Responsibility in particular does not end at the factory gates!

A purely statistical suspicion cannot guarantee a real causality though, but is also not sufficient to attribute a *moral* **guilt,** even if a possibly existing legal regulation of liability takes the damagers into responsibility.

With respect to moral judgement, it seems clear that here causal origination usually cannot be attributed to one single individual agent, nor can causation be solely assigned to one single realm, if development and acceleration are dependent on a multiplicity of mutually interacting and escalating impacts. Participating and contributing individual agents have to bear a certain co-responsibility, not only with respect to different sorts of role and task responsibility, but also in moral and legal terms according to their active, potential, actual or formal participation. For example, in gene biology the exposure of gene-technologically produced bacteria cannot remain only within the moral responsibility of individual scientists, who could not even possibly foresee or predict some harmful side-effects. This is certainly an intriguing example, drastically increasing in prominence today and in the near future. Undoubtedly, all responsibility of researchers in science and technology has to be prevention-oriented and take into account possibly affected individuals, species and ecosystems, whenever and wherever adverse developments and damaging effects can be pre-assessed and avoided. This is true with directly applied technological projects. Preventive responsibility is particularly important in these critical realms, situations and constellations.

With regard to responsibility in general, it is not only corporations and institutions in economics and industry which have to bear responsibility, but also the state and its representative decision makers. Corporate responsibility has to be connected with individual responsibilities of the respective representative decision makers. This is true also for big technology projects, particularly if they are run by the state itself. There should be not only a legal, but also a moral balance of powers in terms of checks and controls similar to the traditional distribution of power between legislature, government and jurisdiction.

Again, the upshot of this in terms of moral responsibility might be formulated thus: the extension of individualistic responsibility is to be combined with the development of a socially proportionate co-responsibility, and with the establishment and analytic, as well as institutional elaboration of corporate responsibility and a new sensitivity of and for moral conscience. Types of responsibility have to be analysed in a more differentiated way than hitherto. Only in this way we may be able to cope with the most complex structures of causal networks and far-ranging consequences of actions, be they individual, strategic, collective, or corporate. Concepts for a more social orientation of responsibility and conscience should be given most attention. Ethics and moral philosophy have to take these new systemic challenges by technically multiplied possibilities and impacts of action and systems networks seriously. This does not mean that our traditional basic intuition of ethics and morality has to be denied. On the contrary, in so far as morality has to deal with the well-being of other men and creatures (maybe even in the sense of Schweitzers' "reverence for life"), analytic and normative ethics have to pay attention to the ramifications of systemic impact and action patterns. Therefore, an extension of our horizon of moral judgement with respect to future generations, quasi-rights of nature, higher animal species etc. is important, if not paramount for any ethical agenda.

Thus, there exists an ethical obligation for humans to take care that especially humankind - as well as other natural kinds dependent on the human power for intervention - does not get extinguished. It is true that individual beings, which have not yet been conceived, have no individual moral or legal right to be born, and one cannot impose an individual obligation on particular human couples to procreate, but it seems to be a sensible extrapolation from the constitutional rights of humankind, which are else often only constructed as rights of repulse and protection, to develop a collective responsibility of today's living humans that they must not let their species be extinguished or destroyed. Humans have not only the, negative, responsibility to leave behind wholesome conditions of environment and life for future generations, which means they should not totally exploit non-regenerable raw materials and should refrain from lethal poisoning, depletion and destruction of the environment. They collectively, also have an obligation and responsibility to actively prevent this from happening and to work for a future existence of humankind in life conditions worthy of human beings. This is, at least, a moral demand which originates in the integrity and continued existence of humankind, which are considered the highest desirable values by various ethical systems. Even a version of Kant's formal Categorical Imperative refers to the actual content of the 'principle of humankind and of any reasoning nature' as things in themselves.

Morally judged, then, future generations' relative rights or quasi-rights to existence do exist, even though no singular existence of a non-conceived individual can be sued for on a moral or legal basis. Thus, certain general human and moral obligations transcend those which are individualistically and juristically concretised. Moral value commitments are more comprehensive and determining than moral or legal individual responsibilities. Morality is more than a singular individual responsibility or obligation.

The necessary extension of individualistic responsibility is to be combined with the development of a socially proportionate co-responsibility, and with the establishment and analytic, as well as institutional elaboration of corporate responsibility and a new sensitivity of and for moral conscience. Types of responsibility have to be analysed in a more differentiated way than hitherto. Only in this way we may be able to cope with the most complex structures of causal networks and far-ranging consequences of actions, be they individual, strategic, collective, or corporate. Concepts for a more social orientation of responsibility and conscience should be given most attention. Ethics and moral philosophy have to take these new systemic challenges by technically multiplied possibilities and impacts of action and systems networks seriously. This does not mean that our traditional basic intuitions of ethics and morality have to be denied. On the contrary, in so far as morality has to deal with the well-being of other human beings and non human creatures, analytic and normative ethics have to pay attention to the ramifications of systemic impact and action patterns.

References

Buchanan, A. 1985. *Ethics, Efficiency, and the Market.* Totowa, New Jersey.

Diederichsen, U. & Scholz, A. 1984. Kausalitats- und beweisprobleme im zivilrechtlichen umweltschutz, *Wirtschaft und Verwaltung*, 23-46.

Hardin, G. 1968. The tragedy of the commons. *Science*, **162**: 1243-1248.

Hart, H.L.A. 1968. *Punishment and Responsibility.* Oxford University Press, Oxford, UK.

Ladd, J. 1975. The ethics of participation. *Nomos*, **16**: 98-125.

Lenk, H. 1975. *Pragmatische Philosophie.* Hamburg.

Lenk, H. 1979. *Pragmatische Vernunft.* Stuttgart.

Lenk, H. 1981a. Herausforderung der ethik durich technologische macht. Zur moralischen problematik des technischen fortschritts. Gesellschaft für rechtspolitik Trier. *Bitburger Ges.*

Lenk, H. 1981b. Herausforderung der Ethik durch technologische Macht. Zur moralischen Problematik des technischen Fortschritts, p. 5-38 in Gesellschaft für Rechtspolitik trier (Ed.): *Bitburger Gespräche.* Jahrbuch, München, 5-38.

Lenk, H. 1982. *Zur Sozialphilosophie der Technik.* Frankfurt a.M.

Lenk, H. 1983. Verantwortung für die natur. *Allgemeine Zeitschrift für Philosophie*, **8**: 1-18.

Lenk, H. 1985. Mitverantwortung ist anteilig zu tragen - auch in der Wissenschaft. In: *Entmoralisierung der Wissenschaften. Ethik der Wissenschaften. Vol. II.* Eds: Baumgartner, H.M. & Staudinger, H. München-Paderborn. pp. 102-109.

Lenk, H. & Ropohl, G. (Eds.) 1987, 1993[2]. *Ethik und Technik.* Stuttgart.

Lenk, H. 1987a. Über verantwortungsbegriffe und das verantwortungsproblem in der technik. In: Lenk, H. & Ropohl, G. (Eds.) 1987, 1993[2]. *Ethik und Technik.* Stuttgart. pp.112-148.

Lenk, H. 1987b. *Zwischen Sozialpsychologie und Sozialphilosophie.* Frankfurt a.M.

Lenk, H. 1989. Types of responsibility. *RSA 2000: Dialogue with the Future.* **11**(2): 1-5.

Lenk, H. 1992. *Zwissen Wissenschaft und Ethik.* Frankfurt a.M.

Lenk, H. 1994. *Macht und Machbarkeit.* Stuttgart.

Lenk, H. 1995. *Knokrete Humanität. Vorlesungen über Verantwortlichkeit und Menschlichkeit.* I. Dr.

Mellema, G. 1985a. Groups, responsibility, and the failure to act. *International Journal for Applied Philosophy*, 57-66.

Mellema, G. 1985b. Shared responsibility and ethical dilutionism. *Australasian Journal of Philosophy*, **63**: 177-187.

Olson, M. 1968. *Die Logik des Kollektiven Handelns.* Tübingen.

Perrow, Ch. 1987. *Normale Katastrohen.* Frankfurt a.M.

Sen, A. 1987. *On Ethics and Economics.* Oxford.

Stehling, F. 1989. *Ökonomische Aspekte des Unweltschutzes - Ökoknomie und Ökologie im Konflikt?* Discussion Paper (Nr. 360). Institut für Wirtschaftstheorie und Operations Research. Universität Karlsruhe.

Tsuru, S. & Weidner, H. (Eds) 1985. *Ein Modell für uns: Die Erfolge der Japanischen Unweltpolitik.* Köln.

Vanberg, V. 1982. *Markt und Organisation.* Tübingen.

Weidner, H. 1985. Bahnbrechende urteile gegen unweltverschmutzer. In: *Ein Modell für uns: Die Erfolge der Japanischen Unweltpolitik.* Eds: Tsuru, S. & Weidner, H. Köln. pp. 92-108.

3   Ecological land development and multidisciplinary research

G. LEIDIG

Introduction

It is evident that a modern and dynamic industrial civilisation, undergoing continual development, generates a constant requirement for political steering and problem-solving which, however, must have a sufficient theoretical basis if it is to be efficient. Consequently, science finds itself confronted with the problem of supplying adequate answers to the constantly emerging and already-known - but as yet unsolved - key problems of the industrial society, e.g. ecologic-economically connected problem areas. Thus the classical individual sciences are frequently out of their depth when faced with problems reaching beyond their own disciplines. As theories gleaned from individual sciences are often in no adequate position to record and explain complex problems in ways concurring with reality, the central theme of the question is, therefore, whether the mono-disciplinary research strategy which has been dominant to date should/must be enhanced by a multidisciplinary nature - also, and particularly, with regard to regional science. For it is the key task of science to provide secure knowledge about the environment and the mechanisms of its functionality. This is being made more difficult at present, as the problem constellations to be solved by science do not adapt themselves to each of the historically grown disciplinary boundaries.   They make it necessary for both scientific disciplines and individual researchers to keep an open mind for questions which go beyond on single subject.

Land development, as a sub-system of regional science, means the dynamic, planned shaping of the relationships between land, property and usage with the aim of creating new system structures. These must, in turn, be orientated around the ecologic-economic functionality of future systems of society so that the subjective legal conditions are in accordance with the objective legal conditions.   If aspects of environmental protection are dominant, it may be referred to as, "Ecological Land Development".

Three theses can be derived from the definition of Ecological Land Development within and it might further be considered that regional science is a moulding process:

1.   of various, interwoven non-linear dynamic systems;
2.   whose key problem areas show that an interwoven interaction of various sciences is necessary;
3.   which only efficiently creates new system structures if it succeeds as rapidly as possible in breaking up antiquated modes of thought and to dispense with "territorial behaviour" specific to any one subject - both in theory and in practice.

This chapter is, therefore, an investigation into whether newer approaches to research possess a suitability profile which opens up the possibility of multidisciplinary co-operation and promotes innovative problem-solving.

Approaches to research

With regard to the question to be discussed in this paper, multidiscipliniarity is the problem-induced integration of knowledge, methods and instruments from other sciences in order to solve complex tasks. Four quite new approaches to research will be analysed below from this perspective:

1.   artificial intelligence research;
2.   artificial life research;
3.   virtual reality research;
4.   chaos research.

Artificial intelligence research

Although this discipline has, over the last few years, lost a little of the euphoria which still surrounded it in the mid-1980s, a promising instrument results from it - expert system technology - which is able to provide constructive and innovative concepts for complex inter-connected problems such as must be methodically solved by regional science. The proposed solutions from expert systems can gradually compete with those of human experts, at least within certain limited areas of knowledge. Such systems possess a database relating to a certain specialist area, a basis of knowledge consisting of coded rules which draws conclusions, and a fast deduction mechanism which systematically applies the rules to the current problem, for example, ecological land development.

Expert systems open up the following spectrum of application:

1. simplification of the provision and administration of comprehensive bases of information and knowledge, e.g. with regard to ecosystem structures, the often interwoven patterns of economic use as well as the legal regulations which must be observed;
2. guarantee of access possibilities to knowledge and experience of experts from different scientific disciplines who are not currently present;
3. improvement of the informational foundation of concepts, e.g. of ecologic-economic land development by means of:

    i.   a locally independent accumulation of knowledge;
    ii. the integration of knowledge from various experts in connection with the necessary adaptation and clarifying of inconsistencies;
    iii. a duplication of knowledge and conservation;
    iv. an eradication of the need to permanently re-analyse/re-evaluate solution concepts.

It follows, therefore, that the application of current expert systems within ecological land development, is primarily aiming at a qualitative improvement of information systems and processes. Thus, specialist knowledge, relevant to a problem, can be both stored on a long-term basis on such systems and be available at all times. The result is that the expert system technology represents a new innovative approach to overcome the interconnected problems which occur.

Artificial life research

In complex systems, e.g. the ecosystems of a landscape, individual elements interact according to such complicated mechanisms that their patterns of behaviour cannot be predicted using standard linear equations. For regional science, therefore, it follows that it makes little sense to base such moulding processes on simplifying approaches. If these are to be efficient, the attempt must be made to analyse and understand the complexity potential which might occur. For this purpose, the young scientific discipline of artificial life research, which emerged in the 1940s, provides promising instruments. In the particular case of land development and the observation of a larger area, it is not merely the concrete question which often leads to a much better understanding of the problem to be solved.

The object of research in the artificial life approach is that life-forms, generated by experts, within the computer develop according to the laws of natural life. The growing understanding of ecological effect mechanisms, in connection with the increasing performance of modern computers, is increasingly placing research in a position to copy the masterpiece of nature, living systems.

Artificial life is devoted to the shaping and research of lifelike organisms and systems created by man. The nature of this material is inorganic, its core is information and computers are the incubators which bring forth these new organisms. Just as medical research has managed to create the processes of life partly in test tubes (in vitro), biologists and computer specialists hope to create life in silicon chips (in silicio).

Artificial life scientists are pondering the possibilities - and the resulting starting points for ecological land development - of generating, developing and observing living systems which are as isomorphic as possible to those in the real environment. There have been attempts to influence the course of evolution for many living systems on the earth and beyond. This great experiment could not only lead to a deeper understanding of life as a whole, but could also provide the possibility to use its mechanisms to assume part of this work. This might even lead to the discovery of powerful laws of nature which not only govern biological, but also every other kind of complex, non-linear, self-organising system.

Artificial life research, which allows a "post-biological" future to be seen on the horizon, could contribute to improving the understanding of ecological systems and thereby expand the spectrum of possibilities, e.g. ecological land development based on simulation calculations. The results will then be analysed by complex instruments, in such a way that is not possible in reality. Based on these results, statements about environmental influences on ecosystems can be derived without really having an adverse effect on them. Consequently this can bee seen as a research-strategical opportunity potential which regional science should not leave unused. However, the prerequisite for this is a willingness for

46

multidisciplinary co-operation.

Virtual reality research

As the world in which man lives is continually increasing in complexity and its individual systems - natural environment, economy, society - are affecting each other to an increasing degree, decisions (particularly those in regional science) are becoming more difficult and have wider-ranging consequences.

In the future, the primary research and problem-solving instrument will most likely be the computer. Although it has, in the past, already fundamentally altered the picture scientists have of material reality, virtual reality research might open up completely new dimensions. Virtual reality (VR) is very practical. It turns the traditional computer - which is utilised today as an appliance with which to process numbers and letters - into a machine which generates its own virtual reality. Scientists and computers are at the threshold of a symbiotic relationship.

VR is an instrument for visualising and modelling which contributes to conceiving plans and developing concepts in order to meet the challenges of dynamically changing systems. VR makes three-dimensional, real-time simulations possible which the researcher, as an interactive participant, can manipulate and alter in many ways.

Apart from the approaches to research already mentioned above - artificial intelligence research and artificial life research - are becoming increasingly interwoven with the virtual reality approach. Thus, not only will the technical possibilities improve but also, in all probability, their spectrum of applications in various scientific systems. Virtual reality research is already providing instruments today which appear to be of use for practical work such as land development. A great advantage of VR is that it enables an exploration of a computer-generated landscape by moving about in it. Instead of sitting in front of a screen or a map, the planner responsible for land development, for example, is integrated into a three-dimensional graphical environment in which this virtually generated landscape can be influenced by certain measures employed in bringing about the development.

With regard to the development and use of VR research for the problem described in this paper, it can safely be said that today's theory is literally tomorrow's reality. Even if the systems are not yet in a status nascendi, it may be assumed that the technology will develop according to an exponential pattern of growth so that any deficits should be eliminated in the medium term. Virtual reality research, will in future, provide regional science/ecological land development with a wide spectrum of application which will contribute towards developing new, innovative concepts. The promising methodical possibilities of VR lie less in a reproduction - as true to nature as possible - of exterior physical reality than in the simplification of navigation and the visualisation of abstract data. In the long term, VR should evolve as one of the most powerful instrument for mastering the flood of data which is becoming ever more complex.

Chaos research

The above approaches to research will bring about the elimination of deficits in models and instruments currently being used. Thus, planning models for the re-development of land and property will, in the future, be shaped more realistically. Current models succeed only partly in recording - with particular emphasis on specific problems - the interwoven nature of various elements and processes which occur in complex systems, with the consequence that the way they model actual structures is frequently in adequate.

Chaos research is also giving innovative impetus. Planning, which precedes development, has the function of preparing future actions. Several alternatives must thereby be strategically considered and held in readiness. In the end, only one of the alternative actions will be carried out. Due to the uncertainty of non-linear systems assumed in chaotic processes, it is not possible to make clear forecasts.

Non-linear models no longer model all the individual causal chains but knots in which feedback loops interconnect. The goal is not a forecast but a means of disrupting the model by varying the parameters, thereby learning something about the critical points of the system and about its capability for resistance.

Analogies of the working mechanisms, from the perspective of ecological land development, show that relationships do exist. This means that the chaos-theoretical approach in land development certainly comes into its own. It was, of course, possible to give plausible reasons for development processes before now. In this way, for example, the necessity for development measures was derived from the ageing process of extremely interwoven systems without the ability to regenerate themselves.

Here, however, the chaos theory opens up new perspectives. For example, crises emerge in points of bifurcation but in their proximity, the systems mostly remain very sensitive. Is it not a land development "crisis" brought about when the sale and exchange of land are not sufficient to bring about the desired order, and measures have to be taken by the state? It is known, from experience, that solutions become possible in processes concerning land development which previously often appeared unimaginable. This is the case for both property and land development and may be explained in terms of the balance of interests. However, this explanation often seems insufficient. Does the system not react to redistribution so sensitively at the point of bifurcation, that old crusts break open to new possibilities? To each decision, made at the branching point, belongs the strengthening of something very tiny by iteration. Strengthening small fluctuations is the lever to creativity.

Through phase locking, i.e. if many individually oscillating systems (= participants in the process) couple, through fixed-phase coupling, out of a state of chaos to a common oscillation unfolding a group rhythm, dreams can create new realities.

The investigation of chaos and its working mechanisms strengthens the initial theory of a holistic approach, e.g. within ecological land development. In this context, the chaos theory provides new patterns of thought and research whose potential should not remain unused.

Conclusion

Based on the above discussion, it is evident that the approaches examined are in a position to give models of regional science (e.g. for land development and innovative impulses concerning their conceptual further development) so that the complexity of reality will be taken into consideration more than in the past.

The prerequisite for this is the willingness to enter into multidisciplinary co-operation. Today, when almost all problems are intertwined with one another, multidiscipliniarity is not only a desideratum of theoretical reason, but a part of a strategy for survival. Every scientific discipline must realise a permanent dialogue and transfer of information with other disciplines which are relevant to the problem in order to be able to make the latest level of research knowledge its own.

The task of this contribution was to indicate which streams of research it makes sense to enter into dialogue with - even if many of these are still in the so-called "teething" phase. But the development of science in modern times is taking a more exponential, rather than linear course, which is why thrusts of innovation - and thereby the accompanying progress of knowledge - contribute to diminishing the faults which can still be recognised.

Regional science must be able, in future, to feel its way into the patterns of thought of other disciplines with a high degree of sensitivity, thereby respecting the ways in which these are different: the colleague from another field is not a projection screen of one's own shortcomings, nor a symbiotic substitute for a part of oneself, nor an object for the egotistical satisfying of one's needs.

In the near future, regional scientific research will be taking place according to completely new schemes. Furthermore, its instruments will be changed both qualitatively and quantitatively.

References

Bullinger, H.-J.,Kern, P. & Braun, M. 1995. Mit der virtuellen Realität in die Zukunft, *Office Management*, 7/8, 12.

Gräfrath, B., Huber, R. & Uhlemann, B. 1991. *Einheit, Interdisziplinarität, Komplementarität*, Berlin & New York.

Hoisl, R. 1994. Eine bodenordnerische Betrachtung zum Chaos. *Vermessungswesen und Raumordnung*, 56(6/7), 318.

Kurzweil, R. 1993. *KI - Das Zeitalter der Künstlichen Intelligenz*. München - Wien.

Lansky, T.F. 1993. Simulation: künstliches Leben. *CHIP 1993*, 6: 44.

Leidig, G. 1983. *Raumplanung als Umweltschutz*. Frankfurt/M., Bern & New York.

Leidig, G. 1984a. *Ökologisch-ökonomische Rechtswissenschaft*. Frankfurt/M., Bern & New York.

Leidig, G. 1984b. Aspekte einer ökologisch-ökonomischen Rechtswissenschaft. *Universitas*, 39(H.7): 759.

Leidig, G, 1990a. Interdisziplinarität und Wissenschaftsentwicklung. In: *Natur und Gesellschaft*. Eds: Klawitter, J., Kümmel, R., & Maier-Rigaud, G., Berlin, Heidelberg, New York, London, Paris, Tokyo, Hong Kong, p. 183.

Leidig, G. 1990b. *Bodenschutz. Gegenstand interdisziplinärer Forschung*, Frankfurt/M., Bern, New York, Paris.

Leidig, G. 1990c. Interdisciplinary research. In: Soziale und ökonomische Aspekte der Bodennutzung. Eds: Fitch, D.B.S. & Pikalo, A., Frankfurt/M., Bern, New York, Paris, p.7.

Leidig, G. 1992. Ökologische Raumplanung als Multidisziplinforschung. *Umwelt- und Planungsrecht*, H8: 294.

Leidig, G. 1993. Expertensystembasierte Bodenschutzpolitik. In: *Bodenpolitik und Infrastruktur*. Eds: De Leeuw, A. & Priemus, H., Frankfurt/M., Berlin, Bern, New York, Paris, Wien. p.223.

Leidig, G. 1995a. Research Strategies in a Time of Change. In: *Gesellschaftssysteme im Umbruch*. Eds: Schinas, G.M., Trappe, P., Frankfurt/M, Berlin, Bern, New York, Paris, Wien. p.45.

Leidig, G. 1995b. Rechtswissenschaft im Wandel: Aspekte eines ökologisch-ökonomischen Ansatzes. In: *Gesellschaftssysteme im Umbruch*. Eds: Schinas, G.M., Trappe, P., Frankfurt/M, Berlin, Bern, New York, Paris, Wien. p.53.

Leidig, G. 1995c. Chaosforschung und Umweltschutz. In: *Social Strategies, Forschungsberichte*. Eds: Leisinger, K.M. & Trappe, P., 4(4), Basel.

Leidig, G. 1997. Rechtsökologische Forschung und Chaostheorie. *Zeitschrift für öffentliches Recht*, 52(1): 127.

Levy, S. 1993. *KL - Künstliches Leben aus dem Computer*. München.

Marsiske, H.-A. 1996. Virtual reality als Medium. *PAGE 1996*, 11: 76.

Mittelstaedt, W. 1997. *Der Chaos-Schock und die Zukunft der Menschheit*. Frankfurt/M., Berlin, Bern, New York, Paris, Wien.

Rheingold, H. 1995. *Virtuelle Welten*. Reinbek bei Hamburg.

Seele, W. 1992. Zur Problematik der Grünen Landneuordnung. In: *Bodenordnung in einem neuen Europa*. Eds: Kocher, G., Schinas, G.M., Frankfurt/M., Berlin, Bern, New York, Paris, Wien. p.245.

Sherman, B. & Judkins, P. 1995. *Virtual Reality*. München.

Spindler, H. 1995. *Die Umwelt und die Zukunft des Menschen. Eine Philosophie der Umwelt*. Hanau.

4 Recent developments in multifunctional rural land development in The Netherlands with respect to legal instruments

J.K.B. SONNENBERG

Introduction

This chapter deals with the recent developments with respect to the legal instruments that serve multifunctional rural land development in The Netherlands. The Land Development Act of 1985 provides a wide variety of instruments for the development of all functions of the rural area. It is recognised that this legislation (especially the legislation with respect to land consolidation and land redevelopment) is applicable to most of the problems in rural land development. On the other hand this legislation is also experienced as being somewhat too complicated. Therefore, a tendency is growing to consider using either the more simple instruments that are provided by the Land Development Act and the Civil Code, or to simplify the specific legislation concerning rural land development.

The Land Development Act 1985

The Land Development Act of 1985 identifies four types of land development, *viz.* land development, land consolidation, land adaptation and land consolidation by agreement (Table 4.1).

Land consolidation is intended for mainly agricultural areas, where other uses are of minor importance. In addition, a few measures can be taken and facilities created for the conservation and development of nature and the landscape, for recreation and for other public uses. Land consolidation involves the reallocation of land in the whole area of a land development project. Reallocation involves the realisation of a new parcellation in which, firstly, the rights of landowners and tenants are established, then the owners' lands are pooled and re-parcelled and finally, the new parcels are allocated to their new owners. It is possible for a project area to be divided into a number of reallocation blocks, which allows the project to be executed in stages. In practice, however, this is rarely done.

Table 4.1 Types of rural land development

Types of rural land development	
According to the legislation	Practical variants
land consolidation	Land consolidation with limited measures
land re-development	Land re-development with limited measures
	Land re-development for special purposes
land adaptation	
land consolidation by agreement	

Land redevelopment is intended for those areas where other functions besides agriculture play an important part. The functions may be existing or planned. The areas may be located near major cities, where land has an agriculture or nature function and is needed, for example, for recreation or road construction. Areas outside the urban sphere of influence may be important to nature and landscape conservation, as well as to agriculture. In areas where different functions exist side by side, land redevelopment is the most appropriate type of land development. The Act provides that land redevelopment may involve reallocation of land in the whole project area or in a part of it. As in land consolidation, the reallocation area can be divided into a number of reallocation blocks.

The main differences between land consolidation and land

53

redevelopment are:

1. land consolidation is intended for predominantly agricultural areas, land redevelopment for areas where multiple functions occur;
2. a decision to carry out a land consolidation project is taken by a ballot among landowners and tenants; a decision to carry out a land redevelopment project is taken by the provincial government;
3. in land consolidation the whole project area is always subject to reallocation and in land redevelopment the project area may be subject to reallocation in its entirety or only a part of it;
4. in land consolidation expropriation is not possible at all. In land redevelopment projects land may be expropriated only for public uses, such as recreation and nature conservation. For this purpose a separate title is included in the Expropriation Act;
5. in land consolidation projects a maximum of 5 per cent may be deducted from the value of the pooled lands before reparcelling and allocation of the new parcels;
6. the land thus deducted may be used for roads and watercourses and related facilities as well as for public uses such as recreation or nature conservation;
7. in land redevelopment projects a maximum of 3 per cent may be deducted, but only for roads, watercourses and related facilities.

Land adaptation is intended to remove constraints which arise as a result of the construction of an infrastructural facility, such as a motorway, a railway or a canal. Recreational areas, military areas and drinking water reservoirs are also regarded as such infrastructural facilities. The facility should be of regional or national importance. Land adaptation takes place in conjunction with the implementation of the infrastructural facility. The scope of land adaptation is usually limited. Reallocation takes place in one block which does not include the infrastructural facility.

Land consolidation by agreement is intended to improve the parcellation of a limited number of owners in a small area. As a rule, no or few works take place. Land consolidation is, in fact, an agreement between all the owners concerned on the redistribution of their land. This type of land development, at present, takes place mainly by way of the exchange of parcels.

New variants of types of land development

The most important types of land development are land redevelopment and land consolidation. Land consolidation is more or less similar to the type of land development that was known under the previous legislation. Land redevelopment is really the new type of land development that was introduced by the Land Development Act and a majority of the projects that are, at present, in the preparation phase belong to this type (Figure 4.1). Based on these two types of land development, some new variants have been developed (see Table 4.1). When the Land Development Act was drafted there was a dispute about the introduction of a simple type of land consolidation. Finally, it was decided not include it, based on the assumption that the legislation allows the execution of more simple projects in a simple way. Based upon current practice, quite a number of simple land consolidation projects, which referred to as reallocation projects or projects with limited infrastructural measures, are in preparation (Figure 4.1) or in execution.

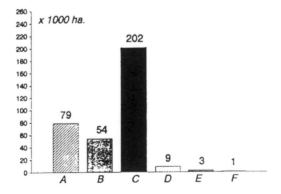

A: land consolidation
B: Land consolidation with limited measures
C: land redevelopment
D: land redevelopment with limited measures
E: land redevelopment for special purposes
F: land adaptation

Figure 4.1 Rural land development projects in preparation phase (1995)

Land consolidation with limited infrastructural measures

In these projects, the main goal is to improve the parcellation of the farms. Only a few measures are taken with respect to the improvement of the infrastructure and the costs of these projects are less than 30% of the average costs of a normal land consolidation project. It has already been proved that these projects can be carried out in a simple way and that the preparation and execution proceeds much faster than normal projects.

Land redevelopment with limited infrastructural measures

As with respect to land consolidation projects, it is assumed that in some special cases it might be possible to carry out land redevelopment projects in a simple way. These special cases may arise when only reallocation of land is needed to redevelop an area and only a few small measures have to be taken to improve the infrastructure or other facilities. For example, this may be the case when there is enough governmental land available for non-agricultural purposes and this land has only to be reallocated. As is illustrated in Figure 4.1, so far only a very limited number of projects fulfil to the specifications of this type of land redevelopment.

Land redevelopment for special purposes

It is assumed that land redevelopment is an adequate instrument for the redevelopment of a particular part of a rural area. In those cases, only the surrounding land should be included in the land redevelopment project and the main goal of the project should be the redevelopment of that specific central part of the area, e.g. to develop a nature reserve. The reallocation of land and other measures, such as improvement of roads and watercourses, nature development and landscaping, may be included in the project as far as these measures serve the particular goal of the land redevelopment project. In this way an attempt must be made to keep these projects as simple as possible. The costs, however, will be about the same as in normal land redevelopment projects. As is shown in Figure 4.1, as yet only a few projects belonging to this type of land redevelopment are in the preparation phase. However, it is expected that in the near future some more projects of this type will be implemented in order to realise the governmental policy with respect to the, so-called, strategic green projects.

Acquisition of land by land development

The Land Development Act provides, in general, two instruments for the acquisition of land; systematic deduction, and expropriation. The legislation with respect to systematic deduction is divided into three different parts (see Figures 4.2 and 4.3):

1. systematic deduction for local roads, watercourses and related facilities such as tree belts;
2. systematic deduction for nature, landscape and recreation;
3. systematic deduction for other non-agricultural purposes.

As is shown in Figures 4.2 and 4.3, a more important instrument for the acquisition of land has proved to be the normal purchase of land. The instrument of reallocation gives the possibility to purchase land over the entire area and to allocate it at almost any desired spot, in order to realise non-agricultural facilities. This is a very powerful facility within reallocation. Figure 4.3 shows that this facility is used even more to-day than a few years ago (Figure 4.2). These figures also illustrate that the use of systematic deduction is decreasing and that of expropriation is increasing.

Alternative instruments and simplification of legislation

To-day, the specific land development legislation is experienced to be somewhat too complicated. Preparation and execution of land development projects take too much time, respectively between 9 and 15 years for an average integrated project, and it is too difficult to anticipate new policies concerning the development of rural areas. Therefore, there is a growing tendency to consider using simpler, alternative instruments that are provided by the Land Development Act or the Civil Code. This could possibly be combined with the use of legislation with respect to spatial planning and, in combination, applied to smaller projects. There is also a growing tendency to simplify the legislation with respect to land development.

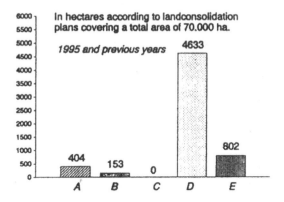

A: systematic deduction for local roads, water courses and related facilities
B: systematic deduction for nature, landscape and recreation
C: systematic deduction for other non-agricultural purposes
D: purchasing or providing land used for nature and landscape by
   reallocation
E: expropriation

Figure 4.2 Land acquisition by rural land development in 1989

Alternative instruments

These can be defined as:

1.   land development without reallocation: expropriation is possible but
        also purchase of land a the right spot is, of course, another
        possibility. To-day, it is easier to purchase land as agriculture is
        less profitable in some areas.   Land redevelopment is still a
        complicated procedure, but avoiding the reallocation procedure
        means a considerable reduction of the complexity of the procedure if
        expropriation can also be avoided.   Alternatively, the experience
        with land consolidation with limited measures in respect to
        infrastructure has proved that the reallocation procedure is not
        necessarily a bottleneck of the whole process. It is obvious that not
        using the instrument of reallocation results in disadvantages with
        respect to the possibilities of purchasing and reallocating of land;

2.   the purchase of land at the right spot and development of this land by using legislation concerning spatial planning;

3.   land consolidation by agreement combined with the use of legislation concerning spatial planning. As land consolidation by agreement includes a simple reallocation procedure, this can be a useful instrument. The disadvantage is that there must be a 100% approval of the land owners and tenants concerned;

4.   an outline plan for a larger area covering the developments of smaller parts by more simple instruments. This provides the possibility for arranging the integration of the separate operations. The establishment of such an outline plan by the government should fit in with the spatial planning activities of the government.

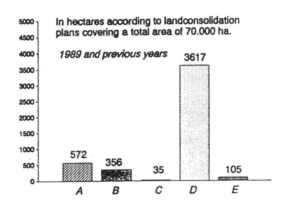

A: systematic deduction for local roads, water courses and related features
B: systematic deduction for nature, landscape and recreation
C: systematic deduction for other non-agricultural purposes
D: purchased or provided land used for nature and landscape by reallocation
E: expropriation

Figure 4.3 Land acquisition by rural development in 1995
Simplification of land development legislation

It is recognised that the Land Development Act is too sophisticated and that a considerable degree of simplification is possible. This would include the following procedures:

1.   removal of the differences between the three complicated types of land development; land consolidation, land redevelopment and land adaptation. There should be only one type that is applicable to all integrated projects, no differences in decision making and systematic deduction of land, reallocation if needed, and acquisition of land for all purposes if needed;
2.   simplification of the reallocation procedure; only valuation of the land that will be exchanged, integration of different procedures for the laying down of documents for public inspection.

During the process of simplifying the legislation, much care will have to be paid to the aspects relating to the protection of the rights of the owner and users of the land.

5 Sustainable development and land consolidation

A. VAN DEN BRINK

Introduction

The planning and layout of rural areas in the Netherlands is geared towards sustainable development. This is evident in government policy on spatial planning, the environment and water, agriculture, nature and the landscape. Social organisations also consider sustainable development an important criterion for desirable changes, while education and research keep up with this trend. A great deal of the research efforts concern questions partially or entirely related to the pursuit of sustainable development. In short, since the publication of the Brundtland report in 1987, sustainable development has received attention from everyone concerned with making and implementing plans for the future of rural areas.

All this attention to sustainable development almost makes it seem that future generations are actually involved in the decisions for which the present generation is responsible. But is this true? As defined in the Brundtland report, the present generation is not permitted to spoil future generations' chances to meet their needs. This seems obvious, but at the same time it evokes numerous questions. What are these future needs? What decisions will future generations take to meet their needs? What will their opportunities to do so be? We do not know the answers to these and other questions, which makes 'sustainable development' a difficult concept.

If it is to be more than an empty phrase and have practical significance, it is important to lend operational content to the pursuit of sustainable development. This has not occurred to a satisfactory extent, at least not in the Netherlands. Many of the political, social and scientific discussions of this issue fall short because they have not been concretely worked out, which results in a Babel-like confusion of tongues. And perhaps the comparison to the ancient Tower of Babel builders is also applicable in the sense that a term such as sustainable development betrays a certain arrogance, as though humankind could shape the future at will.

61

These reflections serve as an overture to deliberations on how the concept of sustainable development can be dealt with in concrete situations where changes in the development of rural areas are at issue. This paper will be limited to, what is known in the Netherlands, as land consolidation and which is distributed over a large number of projects. The purpose of these projects is always to translate general policy objectives into concrete development measures and carry them out. It goes without saying that wherever the theme of sustainable development imbues policy, it should be an important guideline in the implementation of that policy through land consolidation. So the question is how land consolidation can contribute to achieving sustainable development. Before seeking the answer to this question, the concept of sustainable development in should be discussed in more depth.

Sustainable development

The concept of sustainable development springs from the societal conviction that it is time to put an end to deterioration of the environment, nature and the landscape brought about by humankind's economic activities. Developments such as urbanisation, industrialisation and intensification of agricultural production frequently have a negative impact on the physical environment and, consequently, nature and landscape. Countering these negative influences often conflicts with the desire for further economic development. The significance attributed to sustainable development is very much related to how this conflict is dealt with. Accordingly, we can conclude that the pursuit of sustainable development is related to combining conflicting needs, interests and uses.

It is also important to establish that sustainable development is a dynamic concept. Knowledge, norms, values and risk assessments are after all subject to change over time. What is labelled as sustainable development today may be considered something entirely different tomorrow. In other words, sustainable development is not a stable final situation that can be described in absolute, technical terms.

The lack of strictly defined content for sustainable development also makes it clear that it is not a goal in itself, but rather a general principle with which ecological and socio-economic developments can be tested. Key elements are the prevention of irreversible processes, being frugal with base and secondary materials, and working with natural potencies. Against this background, a statement about what can and cannot be considered sustainable development is very much dependent on scale. On a large scale, for example at national or continental level, it makes sense to take sustainable development as the starting point for activities, but a concrete statement about what this should involve is much more difficult to make than on a smaller scale, for example at regional level. In this respect, it is possible to draw inspiration from the famous adage, "Think globally, act locally".

Land consolidation

The question posed above, "how can land consolidation contribute to achieving sustainable development?" can be re-visited. The purpose of land consolidation is to improve the layout of rural areas in accordance with the functions that physical planning policy has assigned to them. This description shows that physical planning plays a significant role. Land consolidation is also closely related to water management and environmental policy. For a long time, the land consolidation instrument had an agricultural basis. In recent years, it has changed course towards giving more attention to attaining objectives in the area of the environment, nature and landscape. A characteristic feature of land consolidation is its project and region-specific approach, in which reassignment of ownership and land use rights occupy a central position. This approach creates favourable opportunities for the integration and co-ordination of public and private interests. In the Netherlands, some 200 projects are being prepared or implemented, accounting for a total surface area of approximately one million hectares (representing 50% of the total surface area of cultivated land).

The contribution of land consolidation to sustainable development can be assessed on three different levels. The first is the regional level, i.e. the land consolidation area as a whole and its immediate surroundings. The second level is the functional uses in the land consolidation area, i.e. agricultural, recreational or natural. Finally, the third level comprises separate land consolidation measures, such as the digging of a ditch, reallocation of land or afforestation. It could be said that measures that bring about an actual change in an area are decisive in terms of the final result. But it should also be realised that those measures cannot be viewed separate from the decisions made on the other two levels. Their interrelatedness is a major factor, and it is considered that this represents common ground between land consolidation and sustainable development.

The region

Some examples may be given to illustrate the concept of thinking and acting on three levels. Firstly, at the regional level, a good example is the Hitland land consolidation project in the immediate vicinity of Rotterdam. This covers a scant 600 hectares, and is a relatively small project in an area that is under considerable urban pressure in the western part of the country. The main objective of the project is to create recreational areas for the urban population. To this end, some farmland will lose its agricultural function, whilst part of the area will continue to be used for agricultural purposes.

The plan is evidently based on a clear view of urban development in and around the area. Nonetheless, the agricultural enclave is vulnerable, because it may be sacrificed to further urbanisation in the long run. Is it still possible in this case to speak of a contribution to sustainable development? It is considered so, for the recreational areas are a stable element in the urban environment. They are large enough to fulfil an important function for the inhabitants of surrounding cities and sufficiently robust to resist new urban developments. In addition, the remaining agricultural land will be developed in such a way that profitable business operations will be feasible within the applicable restrictive environmental provisions. In this way, the needs of the current generation will be satisfied, whilst at the same time allowing future generations to make choices from a different perspective. Whether and when this will occur is now an open question.

Land use

This example demonstrates the importance of working on the basis of a clear spatial structure. Structures provide direction, create order, also when infrastructure, landscape structure and the ecological structure of an area are concerned. This brings me to the second level, land use. This level concerns the mutual co-ordination of functions and how the functions are fulfilled. An ordering principle with which the Netherlands has gained experience in recent years is what is called the framework approach. The crux of this approach is that a distinction is made between functions requiring a long period of development (such as natural functions) and those required to respond rapidly to market/economic and technological developments, such as agricultural functions. This makes a distinction between low-dynamic and high-dynamic forms of land use. The key question is to what extent these forms of land use should remain separated or whether a certain amount of interwovenness is permissible.

A good example is the ecological main structure, whose construction has high priority in the government's nature policy. The basic idea of this ecological main structure, consisting of nature areas, nature development areas and contiguous zones, is that nature requires an area of its own in order to survive and develop. Human activities that obstruct this must be precluded. Of course, the kind of nature concerned must be examined. Vulnerable and rare species call for more protection - and should therefore be separated from human activities - than less vulnerable species. There are also species, such as birds that nest in grassland, that even depend on certain forms of agricultural management. The government is entering into agreements with many farmers in the Netherlands who are prepared to adapt their business operations to this type of agricultural nature conservation.

The framework approach aims at combining low-dynamic functions in a landscape framework of nature areas and plantations. This framework must be stable in terms of location and managed in line with its intended function. There is room in the framework for high-dynamic functions which can develop freely within the set planning and environmental limitations. This contributes to sustainable development in that the functions do not adversely influence each other. Attention must be paid to the mutual relationships and interaction between the functions. The land consolidation measures must be adjusted accordingly, which brings me to the third level.

The measures

In general, it can be said that land consolidation measures must contribute to preservation of the land's natural production capacity in order to achieve sustainable development. Environmental policy sets standards for this. Land consolidation projects are subject to the condition that, on balance, they are not responsible for new environmental impact. Furthermore, the quality of the environment can be improved in many ways, for example by directing water streams in such a way that polluted water is collected outside

65

nature areas and diverted. Other examples are soil remediation and relocation of stock farms to reduce excessive levels of local ammonia.

Creating the conditions for effective and efficient operation and management also contributes to sustainable development. This applies not only to the improvement of agricultural production conditions, but also to management and maintenance of the landscape. In small-scale landscapes, the question is often how this landscape can be preserved in the face of agribusiness's desire to expand. The quality of plantation on parcel boundaries deteriorates due to loss of the original agricultural function. This has a negative effect on the characteristic appearance of the landscape as well as the ecological function that the area can fulfil. Merely preserving what is left does not offer a sustainable solution, neither for agriculture nor for the landscape. An alternative might be to clear part of the plantation and to enlarge another part, turning them into wide wooded strips, which could be managed as separate units. It is not possible to provide a general judgement on this option; it must be considered from area to area. It is interesting that private land users take initiatives to assume joint responsibility for management of landscape elements.

Implementation of measures must be cost-effective, which means that investments must be tested with respect to their contribution to achieving the goals set.

A final point I would like to make is that we should strive towards simplicity in sustainable development. The creation of complex, artificial situations can endanger the durability of the physical environment. In addition, I would like to mention the Roman architect, Vitruvius, who defined the quality of a building 2000 years ago as the inextricable connection between *firmitas* (construction, solidity), *utilitas* (use, function) and *venustas* (beauty, originality). In the broader sense, these are also the conditions for sustainable development. The provisions we make must be solid, functional and aesthetically pleasing.

System approach

Having discussed the three levels separately, I would like to emphasise their cohesion. Sustainable development is very much related to an area's utility on the basis of its natural location, conditions and potencies, to which land consolidation projects pay considerable attention by means of a system approach that lays the foundation for a co-ordination of land use and area characteristics. The system approach provides insight into the biotic and abiotic processes in the area and their relationships to land use. This insight can be used, for example, to find an optimum arrangement of the various types of land use and the related need for water. The result can be sustainable zoning, which is beneficial to the stability of the hydro-ecological system and offers development opportunities for the various functions. The measures to be taken are then adjusted accordingly.

An example of this is the Bethune polder in the Noorderpark land consolidation area. In this low-lying polder, clean groundwater from a nearby chain of hills seeps to the surface. This water is used for the production of drinking water. The polder is rich in nature values. The quality of the surface water, however, is relatively poor due to the agricultural use of the polder and the surrounding area, endangering the nature values and drinking water collection. On the basis of hydrological research, it was decided to designate the polder as a nature area with its own water management as part of the ecological main structure, ensuring drinking water collection in the future. The inflow of polluted water from the surrounding agricultural area is limited by additional hydrological measures. This is combined with measures geared towards improving the quality of agricultural water itself.

Social support

So far, I may have given you the impression that the pursuit of sustainable development is chiefly a technological matter. This is only partly true. Of course, the creation of conditions for sustainable development involves technical issues. It is necessary to have knowledge of a hydrological system in order to take the right water management measures. But apart from this, which I would like to call the technical substructure, is the sociopolitical superstructure, from which the dynamic character of the concept of sustainable development originated. It is easy to see that there is tension between substructure and superstructure. If something is technically sustainable, it does not necessarily follow that it can count on sufficient social support. Farmers' protests against standards for spreading manure over the land illustrate this. The reverse is also true: social support provides no guarantee for sustainability.

This characterises the complex nature of land consolidation projects. The preparation and implementation of a land consolidation project involves co-operation between various authorities and many agencies, each with its own responsibilities and powers. In addition, plans are made in close consultation with owners, users and other interested parties in the area. Their perceptions of sustainable development may differ considerably. Regardless of how the ultimate result is to be judged, the added value of land consolidation is that it leads, at least, to a joint discussion of alternative solutions, that divergent interests are seen in relation to one another, and that knowledge is gained of the biotic and abiotic processes in the area in question.

Conclusion

In the above, I have outlined a picture of the contribution land consolidation makes to achieving sustainable development. This contribution materialises along various lines, as my examples have shown. Important elements are integration and co-ordination of functions, the forming of clear and stable structures and creating conditions for effective and efficient management. The basis for this must be adequate knowledge of the functioning of an area. In addition, there must be sufficient social support for the plans.

Sustainable development is and remains a difficult concept. It needs to be worked out further, both in terms of consolidation and preservation and in terms of development and dynamics. It is important that theory, policy and practice work closely together.

References

H.N. van Lier, C.F. Jaarsma, C.R. Jurgens and A.J. de Buck (editors), 1994. *Sustainable land use planning*, Elsevier, Amsterdam.

6 Public and legally-binding databases for natural resources

G. STOLITZKA and R. MANSBERGER

Requirements for data protecting natural resources

Our natural environment is damaged by abiotic and biotic effects and, especially during the last one hundred years, by human activities. This damage influences the very complex structure of our living space and is very dangerous for our living conditions. Both scientists and politicians, who have to take actions against these dangers, need complete, actual and correct data about the natural resources in order to acquire causal knowledge of the likely and potential damages. Sufficient, correct and current data describing the environment and the chronological development of the environment, will enable scientists to develop models of environmental networks in order to prepare a serious base for decision making. There are many examples of the requirement for such data:

1.   sustainable cultivation of forests in Austria is regulated by laws and this has been extended, in recent years, to the sustainable cultivation of the living space. Within Austria, this is taken to mean the economic use of natural resources, securing of potential areas under cultivation, project planning without the destruction of the environment, and objective planning;

2.   Austria became a member of the European Union at the beginning of 1995 and since this time, projects relating to the integrative development of rural areas have been financially supported by the EU. For this purpose, agricultural planning and development requires up-to-date information about natural resources since the subsidies, granted by the EU, are dependent upon the size of field and the type of crops grown. The need for extensive and current data is not only important for obtaining subsidies from the EU, but is becoming of increasing importance on a national scale. For

example, farming subsidies from the EU will diminish and these will have to be compensated for through national funding, especially in the case of the high alpine regions. This will impose an extra burden upon the whole population of Austria, who will thus require some form of guarantee upon the fairness of the distribution of the national subsidies;

3. the extension of the infrastructure (motorways, roads, railways, power stations, etc.) also have a profound influence upon the environment. Before any project can commence and approval be given, public opinions must be sought. If a co-ordinated programme of data collection already exists, the redundancy of data collection can be avoided and this might be used as evidence to either support or disprove public concern. Providing the data has been collected without bias, it may be offered to groups such as the eco-activists to process and model.

All the examples given above require serious documents for planning, controlling and for decision-making. Most of the required documents (for all 3 applications) are based upon the same data - data describing the natural resources. It makes sense, therefore, to co-ordinate the process of data acquisition, the storing of data and the post processing of data to avoid both redundancies and cost. Using the tools of Geographical Information Systems (GIS), new geometric and especially semantic (thematic) information can be derived from the existing data. The basic requirement for acquiring the correct results is dependent upon the high quality of input data accessed by using public and legally-binding databases.

This document outlines the initial concepts of implementing such a public and legally-binding database for natural resources. The approach adopted is based upon Austrian conditions, taking into consideration the existing infrastructure and existing databases.

Data for protecting natural resources

Data are the capital stock and raw materials of a GIS. The provision and maintenance of this data is an expensive process and the costs involved often amount to several times the combined cost of both the hardware and software used to manage the data. It is also worth noting that often repeated concept, that the quality of the derived information is largely dependent upon the quality of the stored data. This is particularly true when considering a phenomena as complex as natural resources. Here the following aspects have to be considered:

1. the real world is represented as a model. The modelling of ecological and economical relationships, together with social and legal components, within what can loosely be described as reality, can only be achieved using data that is accurate and correct geometrically, thematically and topologically;
2. the costs of both data acquisition and data maintenance increase exponentially with the degree of precision used to record details of the entities;
3. data can have different meanings, differing degrees of importance, and can convey different information to different users (customers).

In collecting data for specific purposes (e.g. for soil protection), some general rules should be taken into consideration:

1. the geometric boundary of the phenomena must be fixed before the data is measured and the thematic data stored as an attribute;
2. data must be acquired in an objective and quantitative manner;
3. data must be collected to a low degree of aggregation. If classes of data are to be merged, it must be done objectively in a subsequent stage using the post-processing tools of a GIS, and not subjectively by interpretation during the process of data acquisition;
4. data acquisition must be verifiable;
5. data must be collected in a manner, and to a format, suited to analysis by a GIS. The geometric description of objects has to be made using the elements of 'point', 'line' and 'area', and the thematic information stored as attributes. The topological relationships must be considered before and during data collection to avoid extensive post-processing.

The thematic description of an object can be achieved in one of two ways. In the first case, the acquisition of a global area, a homogenous area will be demarcated and the thematic content of the area subsequently interpreted. The quality of the results using this method will be dependent upon the homogeneity of the area being assessed. In the second case, poorly defined boundaries (e.g. a forest stand) will influence the quality of the thematic information. As it is difficult to demarcate the precise boundaries of the trees, or tree species, the interpretation is frequently performed in a subjective manner. This is avoided by using a range of statistical techniques, e.g. sample plots within a defined area, to acquire the data. This disadvantage of this second procedure lies in the difficulty of obtaining a continuous surface of information. The approximation of areal data using point data is a serious problem in the GIS technology.

Remarks concerning the terms 'public' and 'legally binding'

Because the requirements for a 'public and legally binding database' are very similar in context to the cadastre and land register, the two terms, 'public' and 'legally binding' are modelled on the legal requirements of the Austrian Land Register. The Land Register operates by:

1. allowing everyone to consult it, both the register of real estates (parcels) and the ownership of parcels, as well as to examine the cadastral map;
2. only permitted certain documents to be examined by those who have a proven public interest in them;
3. permitting access only in the presence of an employee of the public authority;
4. validating the registration of rights, the acquisition and transfer, together with the restriction or removal of civil rights;
5. demonstrating that the documents within the register are both legal and trustworthy.

If this can be related to the concept of public and 'legally-binding databases of natural resources', it would mean that:

1. everyone could consult the data stored, either by visiting regional or local offices, or access the data using modern technologies (e.g. the

WorldWideWeb). This would be the Principle of Publicity;
2. some data would need to be protected and these data could only be accessed by authorised applicants - Data Security;
3. data access must be controlled - Data Protection;
4. data are legally binding and must, therefore, be of a defined, certifiable standard (certified data) - Principle of Trust;
5. general instructions must be defined for further data processing to avoid misleading and false results and manipulations - Principle of Trust;
6. changes to the data can only be implemented by authorised persons.

Organisation of public and legally-binding databases of natural resources

The establishment of such databases demands a widespread spectrum of technical knowledge and has to be implemented by an interdisciplinary group of experts (agrarians, foresters, surveyors, lawyers, civil engineers, landscape architects, regional planning engineers, etc.). The structure of the information must be identified in such a way that the responsibility for all levels and types of data is clearly defined. In specifying the extent and content of the database, the following questions must be addressed:

1. *Where are the data needed?* In general, there is no need for country-wide datasets. In addition, the priority and relative importance of the data will be dependent upon the region.
2. *What data and what degree of detail is required?* The answer to this question depends largely upon the frame condition - 'economy'. Of course it would be an advantage to have a highly detailed information concerning the natural resources within the database, but the cost of acquiring such data increases in a linear manner with the quantity of data and exponentially with the degree of detail and precision. This aspect must also be acknowledged when the maintenance of the database is considered. In general, all data describing the natural environment will cost more, at all stages, since it has a higher coefficient of change (i.e. is more dynamic) than, for example, the topography.

73

3.   *What data is already available on maps or in existing databases?*
       Austria has a great deal of data describing the natural resources.
       The cost of integrating this data into a common database will be
       dependent upon the format and nature of the data. The integration
       within a GIS database of analogue information (usually maps) is
       more time consuming and, hence expensive, than digital data sets.
       Digital data sets are not without their problems, however, since
       there are fundamental differences between topological and non-
       topological formats.

       The establishment of a countrywide public and legally-binding
database is very intensive in terms of personnel, time and cost. In Austria,
both the land register and cadastre have been converted to a digital format,
the real estate database (GDB) and digital cadastre map (DKM),
respectively. Both are suitable systems for the storage of natural resource
data and would, therefore, be suitable starting points for the natural
resources database. In an identical way to both the GDB and DKM, the
data of the natural resources database would need to be stored in a central
office and geo-referenced to the geometric data of the DKM. The storage
and management of the data is referred to as the 'passive part' of the
information system.
       Data acquisition will require a modular approach. This would allow
each module to be examined beforehand for useful effects, and the cost of
data collection must be estimated, together with the costs of maintaining the
data in evidence. If the results of these checks are positive, approval for the
data collection (or the conversion of existing data) can be given. A 'public
assigned person with a special position of trust' will be responsible for
checking the quality and accuracy of the collected and stored data.
       The 'active part' of the system contains the parameters for the post-
processing, analysis and manipulation of the data. Personnel using the data
in the central database will perform this task, together with the merging of
datasets.
       The benefits of having the natural resource data in a public and
legally-binding database, compared with the multiplicity of existing datasets
containing, for example, soil information can be summarised as follows:

1.   application dependent data are collected by responsible experts

using defined criteria. The verifiable collection of data will be performed with the best possible objectivity and with the minimum degree of aggregation;

2. the data will be legally binding;
3. the data will publicly available and accessed by anyone;
4. data stored in a defined, digital format can be accessed using modern communication technology (e.g. the WorldWideWeb);
5. the post-processing, manipulation and analysis of data will be strictly defined by instructions and will, therefore, be verifiable;
6. the proposal for a public and legally-binding database of natural resources is based on the GDB and DKM, both of which are successful in practise.

Data collection, data processing and the visualisation of the data should be organised in small units (ideally to conform to that of the EU's principle of subsidiarity). In Austria, these units could be identical to the units of administration, which will give approximately 200 regional units. A central office will co-ordinate the collection of data and, in a similar manner to the GDB and DKM, to manage the storage of the data concerning the natural resources.

Outlook

The requirements for the graduated students leaving the Universität für Bodenkultur have changed. During the immediate post-war years, the preferred task was of increasing food production, which is now somewhat less important in the current times of overproduction. The new task for responsible people in the field of natural resources (such as politicians, scientists and public servants) is, 'Countrywide ecologically orientated cultivation' within an 'integrated ecological policy for rural development'. The public and legally-binding database is one step in this direction.

In Austria, the implementation of the 'database for natural resources' can probably be realised without extensive changes to the existing laws and by utilising the structure and format of the land register and cadastre. The principle change will refer to the definition of land use parcels, which must be expanded and that the data acquisition for the land use must be laid down in the Constitution.

Users of the data held within the database for natural resources will have to pay for the data. In a similar manner to the GDB and DKM, the earnings for data access will cover the cost of maintaining the database. In general, the methodology discussed in this study is not necessarily specific to conditions within Austria. Thus, the exchange of knowledge concerning experiences regarding the standardisation of data acquisition, data storage and data processing is a very important task for the future.

References

Bundesampt für Eich - und Vermessungswesen, 1986. *Grundstücksdatenbank - Koordinatendataenbank.* Druck und Verlag, Bundesamt für Eich - und Vermessungswesen, Wien.

Dittrich, R. Hrbek, F., Kaluza, W. 1976. *Das österreichische Vermessungsrecht (Vermessungsgesetz samt Vermessungsverordnung, Liegeschaftsteilungsgesetz, Zivitechnikergesetz, Staatsgrenzgesetz und den übrigen einschlägigen Vorschriften).* Manzsche Verlags- und Universitätsbuchhandlung, Wien.

Kusche, W., Mansberger, R., Schneider, W. 1995. GIS-Datenerfassung aus Luftbildern. *Tagungsband der 1. Boku-GIS Userkonferenz.* Universität für Bodenkultur, Wien.

Marent, K.H. 1989. *Grundbuchsrecht - Kommentar zum Grundbuchsgesetz mit Nebengesetzen.* Industrieverlag Peter Linde, Wien.

Stolitzka, G. 1987. Die Stellung des Geodäten bei umweltrelevanten Planungs- und Administrationsaufgaben. *Österreichische Zeitschrift für Vermessungswesen und Photogrammetrie,* **2,** 61-68.

7 Geographical Information Systems and environmental modelling for sustainable development

A. BRIMICOMBE

Introduction

The metaphor of "spaceship earth" grew from the Apollo mission photographs of a small Earth rising above a desolate moonscape. Increasing awareness of such issues as biodiversity and biogeochemical cycles coupled with the technology to detect and monitor environmental impacts have contributed to a "greening" of social attitudes as reflected in consumerism and politics. Whilst global issues, such as warming and ozone depletion, have had a high profile in the emerging debate on sustainable development, the implementation of sustainability through the land use planning process at the local level is producing its own challenges. This paper consequently focuses on the planning implications of Principles 1 & 3 of Agenda 21, on sustainable development in the face of geo-hazards; that is, naturally occurring hazards such as earthquake, volcanic eruption, landslide, subsidence, hurricane, hail and flooding. Special emphasis is given to the role of Geographical Information Systems (GIS) and environmental modelling to mitigate against geo-hazards in the land use planning and development process. To re-iterate from Agenda 21 (United Nations, 1992):

> **Principle 1**: Human beings are at the centre of concerns for sustainable development. They are entitled to a healthy and productive life in harmony with nature.
> **Principle 3**: The right to development must be fulfilled so as to equitably meet developmental and environmental needs of present and future generations.

Geo-hazards and sustainable land use

Principle 1 seems to espouse a romantic view of man in nature, a notion that sits uncomfortably alongside the International Decade for Natural Disaster Reduction (IDNDR). The IDNDR was prompted by the escalation in human vulnerability in terms of heavy loss of life in developing countries and huge economic losses in developed countries (Degg & Ibrahim, 1995). However, insurance losses resulting from man-made catastrophes have remained at a constant level for the period 1970-92, whereas losses resulting from geo-hazards have increased nearly 10-fold (Swiss Reinsurance Company, 1993). Such losses are predicted to increase exponentially over the next couple of decades. Crustal movement, geomorphic processes of landform change and extreme climatic events are naturally occurring phenomena, with correlations of magnitude and frequency that are likely to frustrate attempts at development and occupation of the landscape in harmony with nature.

Geo-hazards are nevertheless characterised by a non-random spatial distribution and can be expected to occur repeatedly in the same locations. Inappropriate development in areas of known or suspected hazard may not immediately have catastrophic consequences but can leave future generations at risk. Furthermore, the cumulative effects of incremental land use change decisions can result in an overall worsening of the magnitude and frequency of some geo-hazards. Thus, the geomorphic response of drainage basin hydrology to urbanisation, resulting in an increase in flood severity, has been recognised for some time (Fox, 1976; Morisawa & La Flure, 1979). Hong Kong, in these respects, makes a good case study and provides important lessons for other rapidly developing areas, where population pressures are forcing the development of marginal land.

Hong Kong case study

Hong Kong comprises of a mainland peninsula and some 260 islands at the mouth of the Pearl River estuary of the Peoples' Republic of China. The territory has a land area of 1,070 square kilometres of which approximately 80% is mountainous terrain. With a population of 6 million and a vibrant economy, the pressures for urban development on the narrow coastal plains and river floodplains is intense. Indeed, sustained urban growth has occurred over the last 30 years resulting in 8 new towns and the well-known feature of the territory, of high-rise development on steep slopes. Half the population lives in high-rise, government housing estates constructed on platforms, either carved into the hillsides by cut and fill or on reclaimed areas of floodplain and coastal mudflat.

The climate of Hong Kong is monsoonal with an average annual rainfall of 2,225mm. During thunderstorms and typhoons, rainfall can reach intensities of 90mm/hr. The steep terrain leads to rapid run-off concentration and flash flooding in the lowland basins. High storm surges along the coast (associated with typhoons) occasionally combine with high rainfall to produce critical floods. Planned new towns, uncontrolled *ad hoc* private development and incremental changes in agricultural land use on the mainland peninsula have substantially modified both run-off characteristics and drainage configuration. These have resulted, over the last decade, in what has now been recognised as a socially unacceptable increase in flood hazard. The United Nations (UN, 1990) has documented the flooding in Hong Kong and an outline of the government's planned mitigatory response is given by Townsend & Bartlett (1992).

Traditionally, of greater concern have been landslide hazards. The steep slopes of, often deeply weathered soils, and high rainfall, over a pronounced summer wet season, combine to produce a terrain in which slope instabilities are a common feature. For example, four days of severe rainfall in May 1982 and another rainstorm in August 1982, resulted in an annual total for that year of more than 1500 landslides (Brand *et al.*, 1984). Such landslides are a normal hillslope process in Hong Kong. Urban development of marginally stable slopes is likely to bring humans in closer proximity to landslide hazard and has indeed been a problem, which has troubled the government since the first British occupation of Hong Kong in 1841. With the increasingly ambitious, high-rise development of the 1960s and 1970s resulting in denser occupation of steeper areas, however, the potential for disaster was augmented. In 1972, a large cutting for a new development sufficiently oversteepened the natural slopes above Kotewall and Po Shan Roads, for a massive landslide to occur during a rainstorm. An adjacent apartment block was swept off its foundations and destroyed, resulting in 55 deaths. Whilst the engineering cause and effect of this disaster was quickly identified, a more incipient danger - a legacy of rapid housing estate construction of the 1960s - remained to be fully recognised. In the hilly terrain, the formation of large platforms for public housing estates was constructed by cutting hills and filling the valleys. Fill slopes were mostly empirically designed and constructed through the process of end tipping loose soil with little compaction. Thus, a legacy was created of slopes susceptible to catastrophic failure through liquefaction as a consequence of leaking water pipes and/or heavy rainfall. Notable disasters were at Sau Mau Ping Estate in 1972 and again in 1976, resulting in a total of 89 deaths.

The mitigatory responses to Hong Kong's landslide problem are summarised in Hansen *et al.*, (1995). Yet after some two decades of cataloguing, monitoring and reconstructing existing slopes, passing new legislation and enforcing strict design criteria, deaths from slope instabilities are a continuing feature of the Hong Kong landscape. To be fair, the majority of these deaths result from the failure of slopes and retaining walls constructed prior to 1976. Thus, current exposure to flood and landslide hazards in Hong Kong derive from the land use planning decisions of the past, imposed on a baseline of naturally occurring geo-hazards. In theory, sufficient is now known of the mechanisms and processes of geo-hazards. Furthermore, adequate technology exists to measure, analyse and simulate these processes for planners to take cognisance of them and to produce sustainable development scenarios that are not detrimental to future generations in terms of exposure these hazards.

GIS in land use planning

The use of GIS in land use planning has seen rapid expansion over the last fifteen years as a consequence of the simultaneous increase in power and reduction in price of computers, coupled with the general availability of commercial software. The current success of GIS rests with its threefold ability to:

1. organise and integrate spatial and attribute data;
2. allow both simple retrieval and complex analyses of data;
3. permit visualisation and communication of spatial information.

It is estimated that some 80% of all information used by local governments and their planning departments are geographical in nature (Steifel, 1987; URISA, 1990). Planning is thus, a form of spatial decision-making (Cohen & Shirley, 1991). Furthermore, it is well recognised that where human activity is enmeshed with the processes of our physical environment, there lie problems, hazards, resources and opportunities (Cooke, 1992). These "environments of concern" (Jones, 1983), driven by public awareness are central to governments' agenda. Given that most of these concerns are spatially and dynamically complex, new technologies such as GIS have emerged to seek solutions at the local, regional and global scale. Thus, Horwood (1980) has stated that

> ..... to be credible, planning must be cast in an information system context.

Klein (1991) summarises the advantages of GIS in the planning process as:

1. a massive increase in the capability for synthesis and analysis;
2. consistency in applying planning principles, criteria and guidelines;
3. empowerment both to planners to access data and to citizens to access planning information.

The heightened tensions between conservation and development at the urban fringe (Selman, 1992), requires effective and accountable strategic planning for which GIS is clearly suited. Scholten & Stillwell (1990) and Worrall (1991) provide a number of case studies of GIS in planning and policy analysis.

GIS and environmental modelling

GIS technology can undoubtedly be an effective tool in the land use planning process. There are, however, a number of approaches a planner can take when considering land development or land use change in areas affected by geo-hazards. This is illustrated in Figure 7.1. Ideally a planning team should be able to evaluate the mix of structural and non-structural measures to achieve sustainability and, in particular, the costs of any measures against the losses that might arise from the effects of geo-hazards for any proposed development scenario. Due consideration must be given to the magnitude and frequency pattern of the geo-hazards concerned. For example, a 1:20 year event may easily be mitigated against, totally and cost-effectively, by technical measures.

81

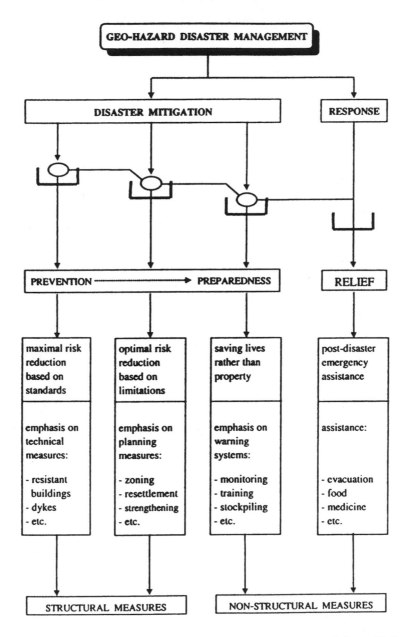

Figure 7.1 Geo-hazard risk reduction measures (adapted from UN, 1990)

However, a 1:200 year event may possibly require a different technical approach or it may be that any structural measures are beyond the financial means available or even are beyond the current engineering state of the art. Thus, it is necessary for a planning team to model geo-hazards in some way in a spatial and temporal context in relation to any land use development programme. Current off-the-shelf GIS technology offers two main approaches to environmental modelling which can be applied to geo-hazards: spatial coexistence and source-pathway characterisation.

Spatial co-existence modelling

The most basic form of environmental modelling is predicated on notions of spatial coexistence (Rejeski, 1993). For geo-hazard mapping in GIS, this takes two predominant forms - event mapping and distributed parameters - which are often used in conjunction with each other (e.g. Siddle *et al.*, 1987).

In the context of landslide hazard mapping, event mapping records all observable failures (usually from aerial photography) and may categorise them according to process (rockfall, debris flow slump) and apparent age. The expectation is that landform units, with a history of landsliding under the prevailing environmental conditions, are likely to pose a threat. Extrapolation of such a threat can either be made by an expert or, less usually, by using statistical techniques (e.g. Carrara *et al.*, 1991), though GIS rarely provide the necessary statistical functionality.

The use of distributed parameters is particularly suited to raster GIS. Here, individual variables (soil, slope, aspect, land cover) thought to be significant in controlling slope stability are mapped over the landscape (e.g. Lopez & Zink, 1991). Such data may be defined in discrete units as in land use categories, or may be handled as continuous data as in slope and aspect. Digital elevation models often play a dominant role. In combining the data to assess hazard, a number of simple GIS techniques can be used including sieve mapping and weighted algebraic combination of layers (e.g. Wang & Unwin, 1992).

A number of problems arise in modelling using spatial coexistence, which may limit its effectiveness. For example, event mapping of landslides and flood extents usually results in a static snapshot of a continuously changing pattern. Thus, verifiable interpretations of magnitude and frequency of events are difficult to arrive at from the data. Many distributed parameter models, used in hazard zonation, are based on assumptions (inductive inferences) regarding the nature and influence of causal factors (van Westen, 1993). The results may be of questionable use. GIS, however, facilitates easy re-runs of such models, with different weightings of the parameters so as to test the sensitivity of the results and thus their reliability.

Source-pathway characterisation modelling

Source-pathway characterisation is based on deterministic models in which an appropriate mathematical simulation of the geo-hazard is used. This assumes that the behaviour of the geo-hazard(s) of interest can be successfully predicted and modelled by a series of equations. Thus, models for geo-hazards such as flooding, strong wind and tidal surge are well developed; landslides and volcanoes less so. However, except for simple models that can be programmed using GIS macro languages, simulation is best performed outside the GIS in a, nevertheless, closely linked loop with spatial data being integrated and pre-processed before being passed as an input to the simulation model. The organisation of the spatial data in the GIS closely follows the structuring of the simulation model. There are two dominant modes - one is again distributed parameters but in the form of grid cells (raster), whilst the other is in the form of a network.

The grid cell approach, models the behaviour of each cell in terms of inputs (energy, water, materials) deriving from both neighbouring cells and from new inputs to the system (e.g. rainfall). The behaviour within the cells (e.g. retention, seepage, overland flow) is derived from distributed parameters, such as soil type and gradient. Outputs are routed to neighbouring cells (e.g. downslope) and accumulated towards the exit of the system (e.g. at the outflow from a drainage basin). This is by far the most popular method and extensively used in hydrological modelling of floods, water quality and non-point pollution in basins (Smith & Brilly, 1992; Gao *et al.*, 1993). Such approaches, thus, usually incorporate digital elevation models (DEM) as the means of modelling the direction and velocity of flow.

The network approach models along known paths of flows using the measured characteristics of those paths (e.g. gradient, channel size). In this approach to flooding, for example, the hydraulic modelling is used to simulate the accumulated flow along a drainage network using true channel details in response to specified rainfall events. It is considered to give a truer simulation of channel capacities and flows than hydrological modelling

(Brimicombe & Bartlett, 1993; 1996). Besides which, by changing channel details, the hydraulic approach can also be used as a design tool in flood alleviation.

Most simulation models are constructed for the general case and inevitably require calibration of their parameters against known historical events. Once a close match is achieved, then possible future scenarios at the local level can be simulated with reasonable confidence. This ability, to either vary the inputs to the model from the GIS and/or alter the model parameters so as to respond to 'what if' type questions, allows GIS and linked simulation models to form spatial decision support systems.

Spatial decision support systems

Spatial decision support systems (SDSS) share a number of common characteristics with the decision support systems, developed by the business data processing community in the 1970s (Densham & Goodchild, 1989). They are principally used to tackle ill- or semi-structured problems, where objectives cannot fully be specified. In Figure 7.1, for example, a very broad objective of geo-hazard disaster management quickly desegregates into sub-objectives, which usually cannot be specified at project inception. SDSS facilitates this by allowing the user to make full use of the data and the models available in a multi-pass approach that allows "what if" type queries to generate multiple solutions. Thus, simulation models linked to GIS, as discussed above, form a strong basis for SDSS. For complex spatial problems, however, SDSS are usually implemented for a specific problem domain. Just as there is no universal GIS data model applicable in every situation, so there is no universal SDSS configuration. Returning then to the Hong Kong case study, in the early 1990s, an SDSS of GIS and hydraulic modelling was set up to tackle the increasingly severe flooding problems of the New Territories. This is a methodology still in use at the time of publication. (Townsend & Bartlett, 1992; Brimicombe, 1992; Brimicombe & Bartlett, 1993; 1996). The structure is given in Figures 7.2 and 7.3. Figure 7.2 summarises the spatial data typically input and integrated within the GIS and the processing of that data prior to input into the hydraulic modelling.

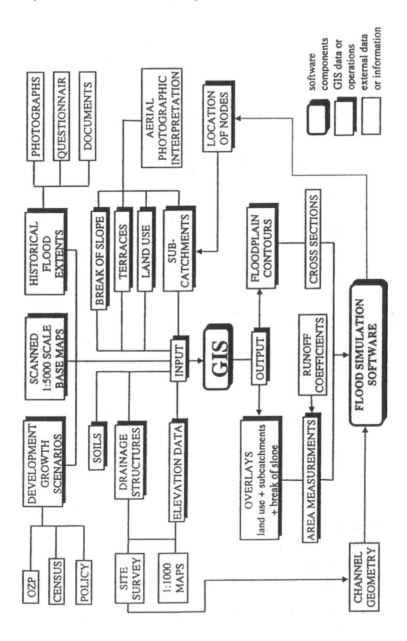

Figure 7.2 GIS data integration and data pre-analysis of inputs to the hydraulic modelling

Figure 7.3 shows the ways in which the outputs from the hydraulic modelling are passed to the GIS and extrapolated into maps, indicating the depth of flood waters over the floodplain from which other maps of hazard and risk can be derived. In the first round, it is necessary to calibrate the model. Current topography, drainage configuration and land use are used to generate the inputs and the rainfall for a known flood event used in the simulation. The results are extrapolated in the GIS, so that the lateral extent of flooding can be visualised and compared with the extent of the known flood event. If data are available for several known flood events, then each of these events may be simulated and compared with reality. Parameters of the modelling, such as the run-off coefficients, are adjusted until the results of simulation for known events have an acceptable fit with reality. It is then possible to pursue a range of "what if" scenarios concerning combinations of possible development plans and mitigatory measures (such as channel enlargement) for a number of different rainfall return periods. Using the multiple outcomes from such an SDSS, it is possible to make judgements regarding the basin management plans which legislate for an approved land use development scenario, incorporating a spectrum of preventive structural measures and prepared non-structural measures. These plans can be physically produced using the map coverages already stored in the GIS. Such an approach should, in theory, be able to facilitate sustainable development in the face of geo-hazards at best, or a "safe" coexistence with geo-hazards within the bounds of possible mitigation, at worst.

GIS, planning and sustainable development

The argument, thus far, has been that living "in harmony with nature", whilst desirable, is a somewhat romantic, if not optimistic, notion in the face of geo-hazards and the numerous catastrophic events induced by earthquakes, volcanic eruptions, hurricanes (typhoons) and so on. Furthermore, geo-hazards exist within a dynamic environment, in which change and variability, albeit in the medium- and long-term, are the norm. Development, however, can itself induce short-term environmental change and lead to increased frequency and intensity of geo-hazards. Thus, to equitably meet the needs of present and future generations, development planning in areas affected by or potentially affected by geo-hazards, can raise a host of possibly intractable issues.

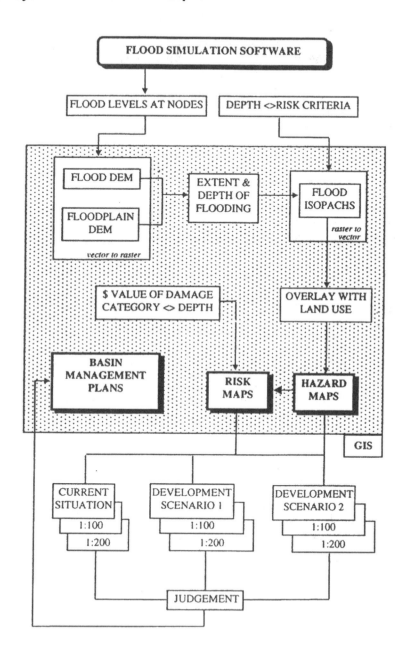

Figure 7.3 GIS post-processing of hydraulic modelling outputs

This chapter has reiterated the arguments of others regarding the advantages of using GIS in the planning process. These are now well established, with widespread use of GIS in physical planning. The use of GIS in environmental modelling has also been discussed and found to have a distinctive role in land use development in the face of geo-hazards, particularly when coupled into a spatial decision support system. Whilst such an approach holds substantial promise, it is by no means a panacea. Constraining issues arise, firstly out of the relationship of land use planning to sustainable development and secondly, out of certain modelling deficiencies in GIS.

It has been a somewhat natural expectation that the existing land use planning system would be able to take up a remit for sustainable development, as environmental concerns are integrated into planning policies. In principle, planners should be able to take account of all aspects of the environment in their work. Indeed, the GIS systems discussed above are one way of doing this. The planning system, however, is not only itself a political process whereby competing demands for land are resolved, but it is also part of a wider political process of reconciling the environment with the economy. Thus

> whilst the planning system is subject to environmental policy, both the system and the policy are shaped and constrained by economic and political forces which may have priorities on a wider stage (Owen, 1994).

The idea that in demand-led economies, environmental resolutions of sustainability will bear fruit through land use planning is therefore naive.

GIS has come under criticism for its positivist approach to the analysis of environmental and planning issues (Lake, 1993; Pickles, 1995). The data modelling assumes that landscapes can be envisioned as mutually exclusive layers, each layer having a reductionist approach to feature classification. Whilst this may be appropriate for physical features with hard boundaries, it is not universally useful. The political processes alluded to in the previous paragraph, are often underscored by cultural landscapes; "culture" here being defined as a system of commonly held beliefs. Points, lines and polygons, even rasters, are poor tools for digitally defining social constructs and cultural landscapes, which are spatially fluid and often have indistinct boundaries. GIS, therefore, fails to handle and analyse important spatial attributes of an area, which can have a dominating influence on planning decisions (Brimicombe & Yeung, 1995).

Notwithstanding these criticisms, the process should not be allowed to atrophy because it is not perfect and cannot be relied upon to always come up with the 'right' sustainable solution. A GIS-based spatial decision support system approach to land use planning provides a means of evaluating a broad range of alternatives and future scenarios. At best, it can indicate sustainable solutions, at worst its outputs can usefully inform the political debate.

References

Brand, E.W., Premchitt, J. & Phillipson, H.B. 1984. Relationship between rainfall and landslides in Hong Kong. *Proceedings 4th International Symposium on Landslides*, Toronto, **1**, 377-384.

Brimicombe, A.J. 1992. Flood risk assessment using spatial decision support systems. *Simulation*, **59**, 379-380.

Brimicombe, A.J. & Bartlett, J. 1993. Spatial decision support in flood hazard and flood risk assessment: a Hong Kong case study. *3rd International Workshop on GIS*, Beijing, **2**, 173-182.

Brimicombe, A.J. & Bartlett, J. 1996. Linking geographic information systems with hydraulic simulation modeling for flood risk assessment: the Hong Kong approach. In: *GIS and Environmental Modelling: Progress and Research Issues*, (eds. Goodchild *et al.*), GIS World Inc., 165-168.

Brimcombe, A.J. & Yeung, D. 1995. An object oriented approach to spatially inexact socio-cultural data. *Proceedings 4th International Conference on Computers in Urban Planning & Urban Management*, Melbourne, **2**, 519-530.

Carrara, A.; Cardinali, M.; Detti, R.; Guzzetti, F.; Pasqui, V. and Reichenbach, P. 1991. GIS techniques and statistical models in evaluating landslide hazard. *Earth Surface Processes and Landforms*, **16**, 172-183.

Cohen, D.J. & Shirley, W.L. 1991. Integrated planning information systems. In: *Geographical Information Systems: Principals and Applications*, (eds Maguire *et al.*), Longman, London, **2**, 297-310.

Cooke, R.U. 1992. Common ground, shared inheritance: research imperatives for environmental geography. *Transactions of the Institute of British Geographers*, NS **17**, 131-151.

Degg, M. & Ibrahim, H. 1995. Geoeducation and the International Decade for Natural Disaster Reduction. *Geoscientist*, **5**(4), 7-8.

Densham, P.J. & Goodchild, M.F. 1989. Spatial decision support systems: a research agenda. *Proceedings LIS/GIS '89*, Orlando, **2**, 707-716.

Fox, H.L. 1976. The urbanizing river: a case study in the Maryland piedmont. In: *Geomorphology and Engineering* (ed. Coates, D.R.), Dowden, Hutchinson & Ross, 245-272.

Gao, X.; Sorooshian, S. & Goodrich, D. 1993. Linkage of a GIS to a distributed rainfall-runoff model. In: *Environmental Modeling with GIS* (eds Goodchild *et al.*), Oxford University Press, New York, 182-187.

Hansen, A; Brimicombe, A.J.; Franks, C.A.M.; Kirk, P.A. & Fung, T. 1995. The application of GIS to landslide hazard assessment in Hong Kong. In: *Geographical Information Systems in Assessing Natural Hazards*, (eds. Carrara & Guzzetti), Kluwer, Netherlands, 273-298.

Horwood, E.M. 1980. Planning information systems: functional approach, evolution and pitfalls. In: *Computers in Local Government Urban and Regional Planning*, (eds Krammer, K. & King, J.), Auerbach, Pennsauken, 1-12.

Jones, D.K.C. 1983. Environments of concern. *Transactions of the Institute of British Geographers*, NS **8**, 429-457.

Lake, R.W. 1993. Planning and applied geography: positivism, ethics and geographic information systems. *Progress in Human Geography*, **17**, 404-413.

Lopez, H.J. & Zink, J.A. 1991. GIS-assisted modelling of soil-induced mass movement hazards: a case study of the upper Coello river basin, Tomila, Columbia. *ITC Journal*, 1991(4), 202-220.

Morisawa, M. & LaFlure E. 1979. Hydraulic geometry, stream equilibrium and urbanization. In: *Adjustments of the Fluvial System* (eds Rhodes, D.D. & Williams, G.P.), George Allen & Unwin, 333-350.

Owen, S. 1994. Land, limits and sustainability: a conceptual framework and some dilemmas for the planning system. *Transactions of the Institute of British Geographers*, NS **19**, 439-456.

Pickles, J (ed.) 1995. *Ground Truth*. Guilford Press, New York.

Scholten, H.J. & Stillwell, J.C.H. 1990. *Geographical Information Systems for Urban and Regional Planning*. Kluwer, Dordrecht.

Selman, P.H. 1992. *Environmental Planning*. Chapman, London.

Siddle, H.J.; Payne, H.R. & Flynn, M.J. 1987. Planning and development in an area susceptible to landslides. In: *Planning and Engineering Geology*, (eds Culshaw *et al.*), Geological Society, London, 247-253.

Smith, M.B. & Brilly, M. 1992. Automated grid element ordering for GIS-based overland flow modeling. *Photogrammetric Engineering & Remote Sensing*, **58**, 579-585.

Steifel, M. 1987. Mapping out the differences between geographic information systems. *The S.Klein Computer Graphics Review*, Fall, 73-87.

Swiss Reinsurance Company 1993. Natural catastrophes and major losses in 1992. *Sigma* 2/93.

Townsend, N.R. and Bartlett, J.M. 1992. Formulation of Basin Management Plans for the northern New Territories of Hong Kong. *Proceedings 3rd International Conference on Floods & Flood Management*, Florence, 39-47.

United Nations 1990. *Urban Flood Loss Prevention and Mitigation.* United Nations, New York.

United Nations 1992. *Earth Summit Agenda 21: The United Nations Programme of Action from Rio.* United Nations Publication E.93.1.11.

United Nations Development Programme 1991. *Manual and Guidelines for Comprehensive Flood Loss Prevention and Management.* United Nations, New York.

URISA 1990. *GIS - Government's Information Solution.* Urban & Regional Information Systems Association, Washington D.C. (video).

van Westen, C.J. 1993. *Applications of Geographic Information Systems to Landslide Hazard Zonation*, ITC Publication No. 15, Enschede.

Wang, S.Q. & Unwin, D.J. 1992. Modelling landslide distribution on loess soils in China: an investigation. *International Journal of Geographical Information Systems*, **6**, 391-405.

Worrall, L. 1991. *Spatial Analysis and Spatial Policy.* Belhaven, London.

8   Institutional systems of agrarian reform

P. TRAPPE

Introduction

Urgent agrarian reforms are currently being considered around the world. However, many of those efforts do not always take their planned course, nor do they always succeed. Numerous examples come to mind. These include many of the former communist states of the Soviet-Union; in Asia (India and Pakistan, in particular); in Africa under comparatively homogeneous conditions (the Sahel Zone is a case in point); in Latin America (especially in Mexico, where agrarian reforms have a long tradition); and last - but not least - there are also the agrarian reforms in Mediterranean Europe, as well as in Central and Northern Europe. In the majority of those examples, the driving force for the planning of agrarian reforms has been the re-distribution of agricultural areas (including the amalgamation of small plots) rather than the 'new' concerns of the preservation and conservation of the countryside and concerns for environmental protection. Thus, the dominant concern of agrarian reforms has largely been driven by the economic infrastructure. This paper will, however focus on the social infrastructure of agrarian reform, which has traditionally been of less interest.

This chapter will examine the typological approach to institutional systems, which has actually been taken into account in three full-scale agrarian reforms. These are the social institutions that are related to the institutional systems, (particularly, when the Western European type of co-operation has recently been introduced) that have to withstand the test of specific institutional pre-conditions of a regional or state order. Furthermore, the so-called 'institutional systems' is frequently embedded in a comprehensive framework (such as a World Trade Organisation) to provide a conditional scope of action. Thus, many of the institutional systems under consideration are multi-stage systems, which must be taken into stringent consideration in the planning of reforms; failing this, there are numerous examples which provide evidence of the inevitable collapse of the system.

Three agrarian reform projects will be considered; in Spain, Northern Tanzania and in the Western Sahel-Zone.

Type 1: Spain - with a focus on the Badojoz plan

In Spain, the objective of agrarian reform was to follow up significant infrastructural measures (such as, water catchment areas, existing irrigation systems, and the pre-conditions for irrigation farming in lowland river valleys), whilst making provisions for the settlement of small-holders. The group of small-holders identified fell into two broad categories; those with existing experience of subsistence farming in arid regions, or those who had previously made a bare living as permanent farmhand or seasonal workers. The institutional system is characterised by exceptionally strict state planning and management (by the Ministry of Agriculture and the Instituto de Colonisacion) and other specialised institutes, i.e. Departments of the Ministry of Agriculture, such as those responsible for hydrology, public works, etc.. For the beneficiaries of the reform, membership of particular co-operatives was declared as a compulsory requirement. Therefore, at this level of the institutional system, compulsory membership of a co-operative was a pre-condition for a successful agrarian reform: it was considered legitimate since it promoted the social integration of the new farmers and demonstrated their ability to act within an economic framework. Models of this kind have employed world-wide in many zones of agrarian reform, in particular to the cotton plantations in the Gheriza scheme, the irrigation zones of Uzbekistan, Turkmenistan and Kazakhstan, and to the compulsory measures of European viticulture, mountain farming and flood-threatened coastal regions.

For over fifty years, the methods of agrarian reform adopted in the Badajoz plan has served to illustrate agrarian reform throughout Spain. The plan served as a very useful pilot scheme and it is significant that the various stages of expansion projected by the Central Planning Committee were achieved. These were; the economic infrastructure measures, land re-distribution measures, the settlement of new farmers in ordinarily new villages, the co-operative infrastructure, tutelage up to the full re-payment of loan for the holding (within 25 years), injunctions for fully irrigated six hectare cultivation, the planning of agro-industries, the establishment of dairy industry, and the promotion of the surrounding, non-irrigated holdings, and their re-structuring (in particular, the fattening of livestock on the basis of fodder grown in the irrigated areas).

This scheme has now been extended to zones at a higher altitude and artificial irrigation has been introduced for larger agricultural enterprises with greater and more diversified production. Despite some setbacks, they have (in principal) been very successful and could form the basis for similar models throughout the Mediterranean region. The trend in development, with regards to consolidating the social infrastructure, lies in maintaining the level of co-operatives. It is also aided by the rise in private enterprise, stimulated by the agro-industries and the latter development towards the formation of larger units within the agricultural sector.

The current situation concerning the remainder of Spain has been less successful, particularly in the south. Here, the agrarian reform and re-forestation programmes are dependent upon a defined schedule of damming rivers and is based upon a defined minimal level of rainfall within the water catchment areas. Clearly, the system works less-well when the rainfall is less than expected. In Andalusia, for example, there has been a quasi-total decline in agricultural production due to extremely low levels of rainfall that has extended for several years. Despite this drought, however, there has still been sufficient water for private homes, for tourism, and for industry. In 1995, the water supplies measured approximately 10% of the capacity of the reservoirs.

Type 2: Dschagga Land, Kilimanjaro

Crop cultivation, under irrigation conditions, exists within the forest belt on the south and east slopes of Kilimanjaro, Northern Tanzania. As in Spain, the Kilimanjaro plan involves a long-term process of change, which has lead to remarkable results in its initial phase. The objective was to introduce farmers of the forest belt to a cash crop of coffee, and to organise its marketing. It was advantageous that infrastructural provisions, in particular irrigation canals, were already established (and had been before the arrival of the Europeans). Furthermore, the social infrastructure was historically well-accustomed to working in groups, such as the abutters' societies and communal self-management on a segmentary basis, long before the arrival of the first Europeans. Following the introduction of the coffee shrub at the Kilimanjaro in 1894, it was possible to utilise the existing social

95

infrastructure for the purpose of the agrarian reform. Initially, marketing was carried out individually, but Farmer's Associations were soon formed on the basis of self-help, and were already firmly institutionalised when the Tanganjikan Bill of Co-operatives was passed in 1932. From this point onwards, a co-operative movement developed, through which the traditional self-help groups and abutters' societies were transferred directly into the legal form of the Western European type of co-operative. These were essentially marketing co-operatives, which besides being engaged in marketing gradually became involved in the purchase of agricultural commodities (utility goods) and consumer goods. The arbutters' societies developed into primary co-operatives, which formed a co-operative union. On Independence Day (1961), the Co-operative Union united some 45,000 coffee farmers. Its main task was to organise the industrial marketing and processing of the coffee cherry, and the selling of raw coffee on the world market, which was auctioned by Union employees. The advantages of the existing social infrastructure meant that no salaries had to be employed in mobilising the farmers, nor did maintenance charges arise for the irrigation systems required throughout the region. 'On the spot' payment by the co-operative for coffee cherries on delivery at its collecting centres, was an essential incentive to the farmers. A further incentive was that both the British trusteeship and the later Tanganjikan Government, granted interim credits and sureties (securities) to secure financing of the scheme.

This favourable development continued until the late 1960s, but was then fundamentally disturbed and changed by a fall in world coffee prices, by the influences of the Ujama policy, and by the Tanzanian government's fiscal policies. Furthermore, self-sufficiency was overtaken by substantial demographic pressures (the population doubled in 25 years) and failed to produce sufficient basic foods. For these reasons, the area of coffee plantations was reduced in favour of other cash crops. The oil price shock of 1973 was a further disadvantage.

Today, farming at the Kilimanjaro is marked by a changing ecological system, largely because an increasing number of subterranean irrigation canals (plastic pipes) have been installed, which affects the growth of tropical vegetation, which requires a high humidity. Similar methods have also, unfortunately, been used in other parts of the world. The Romans, masters of crop cultivation under irrigation, only used subterranean waterways in developed areas and for long distance transport, and never for crop cultivation under irrigation. To compound this problem, British meteorologists have forecast the continual decline of Kilimanjaro's glacier melt water. This is a critical time, since a fall has been predicted in the annual rainfall of the area, whilst the population of the region is increasing - both from an annual 3.2% increase in the birth rate, and due to increased migration into the area.

Type 3: various projects in the Sahel Zone

This section will focus on the zone between Senegal and Lake Chad, which has been considered a veritable experimental laboratory for agrarian reform measures for several decades. The primary objectives of those reforms has been to repair or limit the aftermath of soil exploitation in the southernmost Sahara, and in the transition zone to the damp savannas and rain forests. The situation is marked by the desertification of the Sahara, which is still pushing southwards. This has led to the deterioration of both those soils that had hitherto been farmed, as well as those that had been previously used for cash crops - particularly, in large-scale cultivation, such as peanuts, cotton, and in monocultures. This is also true of soils that had been used for subsistence farming with traditional products (millet, maize and in some cases, rice). The causes of desertification are both climatic and anthropogenic, the latter not only being encountered in large-scale cultivation and traditional hoe culture, but also in traditional pastoral farming (overstocking). Accordingly, the starting point is remarkably diverse although a community-based society, with small-holdings (having a residual nomadic hoe culture in parts) predominates. Permanent settlement leads to the consolidation of land law up to de facto ownership, co-existing with large cash crops and pastoral estates (the latter, for example in northern Nigeria, near Sokoto and Kaduna; the stabilisation of pasturage laws especially in Senegal).

Accordingly, the objectives of agrarian reform are both vague and uncoordinated. They are based on the most disparate institutions, such as traditional forms of co-operation on the basis of permanent or variable land laws, of a modernised co-operative system, of customary law, and also of modern land law introduced during colonisation. Moreover, traditional laws, which are often restricted to local tribes or population groups (e.g. the Fulani), are still effective. Over the past few decades, new supporting bodies such as NGOs and private contractors, which increasingly disregard traditional land orders, have established themselves. Thus, diverse institutions are competing with and rivalling each other.

The social infrastructure of this region has been described by a number of African thinkers as a community-based society. This kind of society continues to be based on a strong segmentary orientation in terms of extended families and clans, in which a group of leaders had already established itself in pre-European times. This leadership exercises certain functions, in particular the regulation and, most of all, the right of use for farmland and for pasturage (contracts between settled and nomadic population groups). In large areas, reform objectives have been reached by means of a consolidated land law, but legal certainty is still insufficient. While the legal systems of the countries concerned all envisage private ownership of farmland, it is not possible to actually carry this into effect. This is a matter of, at least, dual legal systems, i.e. the norm systems that compete with one another on numerous levels. By cultivating perennial crops and implementing irrigation systems, de facto property rights have been established beyond the legal system. In these cases, those involved in cultivation and construction can take effect against third parties, corresponding to continental European property law (on the basis of Roman law). In virtually all countries between Senegal and Chad, de facto property law through annexation may also occur (the fencing of areas to record an exclusive claim to use, often not followed up by actual use). Here too, permanent claims, that correspond increasingly to those protected by property laws, arises particularly following the cultivation of fruit trees.

The reforms are often only of regional significance and relate to specific population groups or particular agricultural products. Privatisation, a solution that occurs in a community-based society, is then usually commenced, leading to market economy behaviour. These efforts have led to the development of numerous private or semi-governmental agro-industries, primarily for the processing of cash crops. These include ginneries, rice mills, tanneries, which ultimately lead to further market economy ventures in the service industries such as the expansion of the hotel industry, of transport societies, banks, etc.. In comparison to the developments characterised in Type 2 (above), this variety of industries is extraordinary but is, however, based explicitly on existing social structures and the ability to put them to good use.

There still remains a host of problems since desertification can, by no means be considered to be over. The migration of the rural populations into the large cities is still in progress. In many cases, the building of dams and the linked irrigation systems has been deferred to prevent the interference with water supplies for the production of energy for the large cities of the south. The large dams of Kossou, in the Central Ivory Coast, and at the lower stretches of the Volta river are given priority over irrigation measures in the northern parts of the country. Although there are contracts under International law for rivers crossing national boundaries, they are hardly worth considering.

Conclusion

An institution can be defined as a lasting pattern of human relationships that have either been consciously shaped or accidentally stable. It can either be exhorted in a society or supported, and actually lived on the basis of generally legitimate conceptions of order. The institutional systems of the three briefly characterised zones of agrarian reform discussed above can be summarised below.

Type 1. The regional developments have taken place under strict central planning, with prescribed and newly established institutions due to the absence of an existing social infrastructure that could have been made use of. Any existing social infrastructure would not have been very effective since the settlers had different origins and came from different social classes. This is a complex system, but nevertheless had clear planning objectives. In addition, within its programme of agrarian reform, it had provision for a central and supra-regional economic infrastructure, irrigation systems, central planning for new settlements, schools, and agricultural co-operatives as multi-purpose societies, together with financial aid for agro-industries.

Type 2. A strong and firmly established traditional institution (autochthonous societies, pre-co-operatives, working groups, self-help organisations) was already in existence in this region. This resulted in an extremely favourable starting position for Western European types of co-operatives, with clear and comprehensive economic objectives for co-operative activities. This resulted in favourable effects on further reform objectives, beyond the actual process of agrarian reform (i.e. the attainment of an agro-industrial level and its co-operative management). The overall

99

result was a general improvement in the standard of living across society at large, i.e. a guaranteed, broad effect of the development intervention. Subsequent problems, caused by burdens on the environment, in the context of fiscal policies, population increases and climactic changes in the Kilimanjaro settlement area have resulted in the efficacy of the agrarian reform.

Type 3. In this example, the programme of agrarian reform sought heterogeneous objectives according to local needs and incentives. This has resulted in an isolated realisation of the objectives resulting from a lack of comprehensive planning and co-ordination. The process had largely been borne by traditional institutions, communities and working groups and there were few developed forms of regional co-operation. The social communities suffered and were largely helpless in the face of the disastrous effects on the environment, such as desertification, oversalting, evaporation, and seepage of dammed rain water.

The key problem to all three types of agrarian reform are essentially the environmental hazards caused by climactic and anthropogenic influences. The latter result from, or are significantly simplified by, regional and supra-regional causes. In these situations, regional planning, if not supra-regional planning with explicit possibilities of control and intervention, seem a pre-requisite to agrarian reform.

9 The necessary foundations of land reform: some painful post-Soviet experiences

C. ARNISON

Introduction

When, some eight years ago, the Soviet Union ceased to exercise a single and central control over its own economy as well as over those of its satellite states in the Comecon area, western observers assumed that amongst the earliest reforms would be those affecting the ownership and control of land. Such assumptions were themselves based on other assumptions, namely; that 'the people' in countries which had been annexed by the Soviet Union after the 1939-45 War would demand restitution of land confiscated from them by the State, and private ownership of land was the single and essential key to the formation of the private enterprises which would be needed to fill the vacuum left by the collapse of centralised production and distribution.

There is now some evidence that neither of these assumptions is fully justified, and that any agenda for the transition to a market economy must include some essential preliminaries which have not yet occurred in the Former Soviet Union (FSU).

Restitution of private land

Political pressure to restore land to its previous owners was indeed irresistible in the Central and Eastern European states and all of them set up procedures to identify and verify land claims. However, the consequences of a successful claim differ considerably from Estonia in the north to Bulgaria in the extreme south. It is not the purpose of this paper to catalogue all the details of land restitution in the ten countries concerned. Certain features, however, are worthy of note, especially in contrast with the experience of the new republics of the FSU.

1.   Restitution of dwellings to their occupying former owners has presented little legislative or ideological difficulty. In most cases it has been rapidly followed by granting new rights to the remaining occupiers by converting tenancies into ownership. Almost all the dwellings concerned in the latter case are apartments, as are many of the former. However, the ownership rights now held are in all cases limited, in some cases quite severely. Indeed, in economic terms the householder's newly acquired financial asset is in most cases potential rather than actual: it usually cannot be sold for some years; there are difficulties in creating new private tenancies; mortgages are always difficult and sometimes impossible. Thus the stimulus to economic activity and to private initiative is greatly reduced. It is even further reduced by severe problems of management and maintenance of the apartment buildings in which a very high proportion of new owners live, making it extremely difficult for owners or prospective tenants or buyers to estimate occupation costs and thereby be able to assess value with any confidence.

2.   Restitution of agricultural land has presented both practical and policy problems. In the case of land which used to be farmed but is now part of an industrial complex or the site of a school, actual restitution is obviously impossible and the former owner is in most cases offered bonds or vouchers equal to the present agricultural value of the land. In the absence of any evidence of market values the valuation of such land is based on notional values having regard to location and soil quality (Rydval & Pesl, 1998) describe the typical case of such valuations in the Czech Republic. But where the land is still in agricultural use as part of a state or collective farm an interesting policy question arises: should a 50-year old metalworker living in an industrial town, with no knowledge of farming, be given, say, ten hectares of land twenty five kilometres away just because he was born there and his family owned it in 1945? Quite apart from the difficulty of giving him access to it if it is now in the middle of a 200-hectare cornfield, the following consequences raise issues which are not easily answered;

   •   if he becomes a farmer his metalworking skills are lost to the industrial economy

- the transfer of his family to the country imposes a new social cost as regards health and education on the rural community
- he is very unlikely to be an efficient farmer and the productivity of his ten hectares will therefore be less than if it had continued to be part of a larger enterprise
- if he leases the land to someone else his motivation to work in the factory is reduced because he now has some unearned income
- if he sells the land he may be the victim of speculators or be cheated.

All these questions have in fact been asked in the author's presence by public officials in the FSU who are well aware of how puzzled their former comrades in eastern Europe have been by these very points, and others. Some consequences of attempts by government agencies, in the person of their officials, to incorporate solutions to such problems in new legislation are examined below.

3.   Restitution of ownership of urban property such as shops, offices, guest houses etcetera is relatively straightforward in principle as surviving legal papers and public records usually enable accurate identification to be made. But the fear of speculative dealings in such property is widespread, coupled with deep concern at the possibility of 'windfall' gains to such owners from the redevelopment of urban property and from the inevitable growth of commercial activity as the new market economies grow.

The sale of such restituted property is therefore prohibited by most former communist governments, for up to ten years.

These problems are well understood by politicians and adminstrators in the FSU states, although in their own cases the political pressures are somewhat different. The chief difference is that there is in effect no need for a restitution programme - certainly there is no general call for it amongst the people. It seems that two whole generations of dispossession is sufficient to erase both the desire for and the evidence of private title to land or buildings. The Soviet Union was in fact complete by 1922, three generations ago, but it must be remembered that the policy of universal large communal farms was imposed over a thirty year period starting in the 1930s, and that many private peasant farms existed right up

103

to the late 1950s. Some landowners, even in Russia itself, only lost their land less than forty years ago, at the same time as did the Estonians, Ukrainians, Albanians and some Georgians, Uzbeks, Tajiks etc. Nevertheless it is the author's experience that in the Central Asian and Caucasus states there is no real pressure for former private land to be returned to the heirs of its owners.

Distribution of public lands

Privatisation, the transfer of publicly owned property to individual citizens or legal persons (i.e. companies), is quite another matter. The rationale for privatisation is not that justice demands it; that outright confiscation without compensation must be reversed. It is that economic theory and practise show that it is more efficient: that greater production will ensue, and further that essential innovation and invention will occur. The cynic might say that this is merely exchanging a socialist dogma for a capitalist dogma; that there are plenty of examples around the world of market economies with private ownership of land and other property which are, or were, very inefficient. The absentee landlord can have quite the same demotivating effect as the anonymous state farm, and the extreme fragmentation of individual land holdings in, say, nineteenth century Ireland or twentieth century Egypt is even more inefficient than the typical Soviet collective farm. Nevertheless, the dogma is almost universally accepted by western advisers and by the all-important donors of aid and funders of projects such as the World Bank, the European Union, the European Bank of Reconstruction and Development, the Asian Development Bank and others.

Putting privatisation in place has proved to be much more difficult than expected. The difficulties appear to fall into three groups: first, who among the population of a new nation should have private ownership conferred on him and how should he be selected?; second, what limits must be placed on his ability to enjoy and exploit his good fortune?; and third, how does one implement the decisions reached? The first two issues are political and social rather than juridical, and many of the new parliaments have debated them for many months, with varying results. We are not here concerned with the content of those debates but with the third group of problems; the implementation of the final (or, more often, interim) decisions, and with the laws, decrees and resolutions through which governments both stimulate and control action.

The legal maze

Much less has been written about the legal aspects of the transition to market economies of the 20-odd new nations of the old Communist Bloc than about the economic ones. For example, when the World Bank held a major international Symposium on the Agricultural Transition in Central and Eastern Europe and the Former USSR in Hungary in 1990, only one of eighteen papers presented to the Symposium was devoted to legal land rights, with four others containing limited reference to them (Braverman et al. 1993). Understandably, the first and central concern of such a Symposium and of those who took part in it, was the economics of the process of reform. Legal matters are seen as the administrative drudgery which follows the higher level, creative work of getting the principles right. Social considerations follow next, being also concerned with principles, and much has been written and said on the effect on the ordinary citizen of unemployment, high inflation and the removal of the social security safety-net which the collective, whether farm or factory, used to provide for all its members. One has to look at the reports produced very recently by or for investors, for example by the European Bank of Reconstruction and Development, (EBRD 1995), to find detailed examination of legal matters.

But the legal drudgery has to be done, and done well. Badly drafted laws simply don't work; inadequate institutional structures - courts, tribunals, registries - and insufficient trained personnel at least hamper and at worst actually prevent the decisions of a parliament from being implemented. As the time of debate passes to a time of decision and thence to the task of giving effect to decisions observers of the process are finding, as I have, that deficiencies in the apparently mundane business of drafting and introducing regulations can have just as direct an effect on daily life in the FSU as the lack of machinery parts and of money to repair roads and buildings. Such effects are well illustrated by the following case studies.

Apartments in Tbilisi, Georgia

Tbilisi is a city of 1.2 million inhabitants and is the capital of the Republic of Georgia, which has 5 million population and territory about the size of Ireland. Most of Tbilisi's citizens live in apartment buildings constructed between 1950 and 1980. On the northern edge of the city is the district of Didi Digomi where 4,900 apartments in multi-storey blocks are administered by a housing manager with one assistant and a staff lawyer. During a visit in June 1995 the author was informed by the housing manager of the estate that over 80% of the residents had exercised their right to be granted ownership of their apartment under the Law on Ownership passed by the Georgian Parliament in 1993. Under this law the new owners of apartments (that is to say, the former occupiers who have now been granted 'ownership') must pay an annual service charge to cover the upkeep of the building. For reasons which were not explained, but which must include the need to avoid inflationary price increases wherever possible, the charge is at present the equivalent of US$1 a year. As a result there is no money with which to carry out even essential repairs, let alone set aside a fund for the renewal of lifts, heating plant, windows etc. Such repairs as are done are paid for by the residents themselves, but the payment is often a mixture of barter and money, and not all residents contribute equally. The result is rapidly accelerating deterioration of such buildings and their services, and complete uncertainty as to the availability of heating and lighting, and even of water, during the winter months which in Georgia are very cold. This means that such apartments have a very low value as potential buyers must either make provision for the extra costs of their own generator and additional water tanks or must face very uncomfortable living conditions. Similar apartments were being offered for sale at $5,000 by their new owners, when the current (1995) rebuilding cost would be at least $20,000. Thus, the result of the failure of the legislators to grasp the necessity of on the one hand allocating repair and maintenance responsibilities to a 'landlord' - most likely in the form of a management company in which the freehold ownership is vested - and on the other hand enabling it to charge a realistic service charge on all lessee-owners equally is to throw 75% of the potential value of the apartment away and thus cut down greatly the amount of new wealth that can circulate round in the economy. I was informed by the Housing Manager that only 12 sales of privately owned apartments had been recorded, out of 4,900, during the first half of 1995. The benefit of the distribution of such property is thus largely lost both to individuals and to the economy.

Private farm, Gorki, near Moscow

About 20 miles south west of Moscow is the village of Gorki, in effect the administrative and residential centre of Gorki Collective Farm. Under the Law on Privatisation passed by the parliament of the Russian Federation in

1993 members of a Collective Farm may apply to exchange the shares in the farm which they received as a result of the privatisation of state and collective enterprises for up to 50 hectares of land from the farm which becomes their private property to farm as they wish. Two years ago one of the former managers of the farm made such an application on behalf of a group of ten members of the collective and was granted 20 hectares of level arable land adjoining one of the farm roads and near to a public highway. The group now 'own' the land in partnership but cannot sell it on the open market. If one of them wishes to sell he must first offer his share to the others and if they do not want it to another member of the collective. The value of the share is a matter of negotiation. In effect, the 'ownership' which is given under the new law is hardly recognisable as ownership in free market terms. They did not choose the land they received either; it was selected by the Collective Farm management committee and was priced according to a centrally determined formula taking account of its location and fertility. They will own any buildings on the land outright as in Russian law buildings are separate from the land they stand on, but they cannot raise money to erect the storage building they badly need chiefly because they cannot use the land itself as security for the loan.

New 'homesteads', Ashgabat, Turkmenistan

Turkmenistan has an abundance of land but a shortage of water. Its collective and state farms are huge in extent - 100,000 hectares is not unusual, but only a small proportion of each, say 8 to 10 thousand hectares, may be capable of any kind of agriculture nearly all of which will be dependent on irrigation from the 1,000 km Karakum Canal. Such an abundance of land permits an apparently generous policy on granting private ownership rights. The Law on Granting Ownership of Land requires each collective farm to identify an area of virgin land within its boundary and to donate it to a pool, which is administered by local municipal councils. Individuals or families who want to set up as independent farmers may apply for the grant of such lands, up to 50 hectares, and their applications are assessed by a committee on the basis of their financial resources, their agricultural education and experience and the number of jobs their enterprise will create. A grant of such virgin land can be for a term of up to 50 years, though it appears that 20 years is favoured, and is free of land taxes for the first ten years. As with all other private land in Turkmenistan, the land cannot be sold for ten years and cannot be transferred out of agricultural use. In the summer of 1994, one year after the Law had been passed, no actual grants could be identified, although it was claimed by officials in the Ministry of Agriculture and Food Industry that a number had been approved in principle.

The site chosen for the first such land grants in the Gawers district, to the east of Ashgabat city, was visited during the Autumn of 1994 and found to be 30 kilometres away from the farm/village centre and five kilometres from a metalled road in an area of semi-nomadic herdsmen where water has to be pumped from wells. Some arable cultivation was evident nearby but the economics of settled arable farming in such a remote location must be highly uncertain.

Just as in the Russian Federation, the absence of a freely transactable land title and the uncertainty of the supply of water makes it almost impossible for such farmers to raise capital for machinery or buildings or for better wells and irrigation works. The value of ownership under such conditions is very low. Yet again the potential benefits of privatisation are largely lost to the economy as well as to the individual.

Other common features of FSU land reforms

In all three countries decisions about who was to receive land, and the particular location of such land, was initially taken at the 'village' level by Committees led by the leaders of either the community or the collective or state enterprise which owned the land. Such bodies and individuals have little incentive to be objective and impartial in their decisions, or indeed to act with any urgency in dealing with applications received. These local decisions must then be approved by the next higher tier of public administration, and finally by the relevant ministry, usually the Ministry of Agriculture.

An excellent summary of the problems, which are being encountered in practice was given by Andonov & Risov (1999). Describing their experiences of the Municipal Land Commissions in Bulgaria, they stated

> During the past five years processes for documenting ownership of land were found to be different in kind and character. In many places there are no documents at all.... Sometimes they are contradictory or invalid. Sometimes pressure is exerted over the work of the Commisssions' staff

by the executive or political powers, interested bodies and institutions. In other cases incompetent interpretation of the law and subjective and unrealistic recommendations cause confusion. Furthermore the different bases for defining ownership in the various legislative acts '...continue to impede land reform.' (Andonov & Rizov, 1999).

The consequences

We can now begin to see one at least of the unforeseen consequences and it is not a consequence of the reform itself but of the preliminary process in which legislators and jurists attempted to translate the political and in most cases constitutional obligation to give land ownership to ordinary citizens into practical law. But we should understand I think that those features of the legislation which western advisers such as myself have regarded as defects were, in most cases at least, intended to avoid even worse defects; worse at least in the eyes of those who did the drafting. Thus the moratorium on resale in the open market for up to ten years which in effect prevents the real value of the new property rights being in any sensible way realised was intended to avoid the worse evils of both speculation and undeserved windfall profits. Several senior officials have said to me that it would be both morally and economically indefensible for a person who had acquired property ownership rights, whether by restitution or by allocation, at no cost to himself or his family, to be able later to sell those rights at a profit. We should not be surprised at such attitudes which are very similar indeed to those of nineteenth century English landlords (Arnison 1984). Their explanations as to precisely why such an event would be unacceptable revealed first a strong ideological belief that wealth must be earned through some direct activity, whether rational or manual, and second that the possibility of directly taxing such gains in order that the benefit of community-generated value passed to the community had not occurred to them.

It may be that the particular conversations I have had in Turkmenistan, Georgia and Russia in the course of four relatively short visits were not typical. But the fact remains that that is the way the laws have been drafted in most parts of the FSU and that no other explanation was offered to me at all.

Another major difficulty is less clearly illustrated by the examples given above. It derives from the answers which most of the draftsmen devised to the vexing question of "who should be the recipients of this wonderful bounty"? Throughout the FSU and its former satellites virtually all state and collective farms are, nominally at least, being privatised. Setting aside those parts of them, which might have to be returned to former owners there are enormous areas capable of division into small farms. Who are to be the new small farmers? In most of these countries the government body dealing with the privatisation of agricultural land is the Ministry of Agriculture whose officials, perhaps understandably, have taken it for granted that the new owners must be able to demonstrate that they are going to be competent farmers, as if the objective of the reform was purely economic; to increase agricultural efficiency, rather than simply political; to give resources directly to individuals. Thus, most of the privatisation procedures require the applicant for a private allocation of land to demonstrate both agricultural competence and the financial ability to undertake 'small farming'. As stated above the powers to propose such allocations are delegated down to the very local level with the result that the few small farms I have myself seen had been given to former managers, administrators or officials of state or collective farms.

Such allocations, incidentally, can be transferred by inheritance even during the initial period when they cannot be sold openly, and this is entirely consistent with a third unexpected aspect of land privatisation; the designation of the family as the proprietary unit of rural land, rather than the individual. This is particularly so in Georgia where the new law on ownership specifies that agricultural land may be allocated only to families and that it is the whole family which is the owner, although one member of the family will be deemed in law to be the trustee for the others. It may be wrong to assume that Georgian families will behave in the same way as the western families with which I am more familiar, but I cannot myself imagine a more potent mechanism for entirely frustrating the formation of an open market in viable agricultural units.

But there is a fourth difficulty which in my opinion will prove to be the most important of all. Its origin is neither juridical, nor economic, nor political. It is a problem of social psychology which is well understood in the West but apparently ignored in the East: market economies require the players in the economy to engage in a continuous process of decision-making, of risk-taking, of arbitrage. Such decisions and actions are, in the West, based on market knowledge, on information gained in part from

110

private sources but in the main from published information in journals and bulletins. Such information is very frequently supplemented by professional or commercial advice from experts or consultants. Without such information and advice fewer decisions will be made because many people will be unwilling to take the risk of acting in ignorance, and many of the decisions that are made will result in loss or failure. Thus the forward momentum of the new market economy will be greatly slowed.

My own limited experience of three FSU states revealed an almost complete absence of either published information about costs, prices or market volume and an equal absence of the services of professional advisers such as lawyers, valuers, accountants, architects, surveyors and engineers. The result is that such entrepreneurial decisions as are being taken are based on uninformed guesswork.

The solution

To a visitor such as myself there appear to be two simple reasons for all the difficulties I have recited. The first is on the surface somewhat masked by the specific problems resulting from the grant of the powers to implement privatisation legislation to an inappropriate ministry with a clear vested interest in maintaining large parts of the status quo. But the first underlying problem is simply the result of an oversimplified approach, both politically and juridically, to the process itself. What is required first is not a law on ownership but a suite of laws, enacted together and drafted in a manner which carefully thinks through their interaction and interdependence, covering ownership, the compulsory registration of titles and transactions, the principles of taxation of both annual and capital values, and a law covering land use with its attendant controls on change of use and development. Many commentators on the post-Soviet scene, including those with far more experience than I have (e.g. Thiesenhusen 1994, Brooks 1994, Csaki 1994), have remarked on the piecemeal approach to land reform legislation and on the operational difficulties created by the absence, for example, of a law on mortgage at the time when the first laws on ownership were passed.

The second, more important reason for many of the difficulties I have described is clear: one of the most important resources that the FSU does not have, and its old satellites have only in relatively small measure, is the professional infrastructure of lawyers, accountants, valuers, surveyors, engineers and public administrators who would have been able to give practical, objective and informed advice on these matters at a very early stage. On this point I take a different view from that of my former colleague Dr. Munro-Faure who has suggested that there are "very well established professional associations in the former Soviet world" (Munro-Faure 1994). There are indeed many associations of architects, doctors, engineers, surveyors etc., but they are not 'professional' in the sense of being guardians of a code of ethical conduct designed to protect the public interest. Indeed, one of the unexpected outcomes of my own few visits is a growing conviction that the presence in our own UK Parliament of lawyers, accountants, doctors and surveyors is on balance very valuable. At least it means that Parliamentary committees and debates are informed at an early stage of most of the things that could go wrong with purely politically drafted measures. But of course it goes much further than Parliament. The citizens and other legal entities (companies and corporations) in the FSU have had to date no way of getting advice on how best to present a case, on how to resist what they may feel to be arbitrary or even corrupt decisions by public officials or on how the rules and regulations will be applied by the courts. Nor, as stated above, have they had access to the broad market intelligence which is offered on every bookstall in the West, and by every valuer, architect and engineer with a private office.

So the necessary foundation referred to in the title to this paper is the presence in society of legal expertise coupled with other professional and administrative skills which are together made equally available to both the state, in the form of legislative draftsmen and advisers who are qualified in other relevant professions, and to the public in the form of accessible and affordable advisers whose integrity is assured by a system of proper training and licencing. The scale of the task of providing the providers of such advice, in terms of both cost and time, should not be underestimated. It will take years to achieve and will cost many millions of dollars in training and education programmes. Implanting - or rather transplanting, for they exist elsewhere and have been successfully transplanted in the past - true professional attitudes and structures in the form of independent and largely self-regulating associations backed by State controls needs long-term collaborative projects such as those currently beginning in Bulgaria, Hungary and Poland. A university-level postgraduate course in Property Valuation is at present being established in Sofia and a postgraduate course

Valuation is at present being established in Sofia and a postgraduate course in Budapest, run jointly by the University of Nottingham Trent, UK, and the University of Budapest, is now recognised by the UK Royal Institution of Chartered Surveyors as of a fully professional standard. In Bulgaria there is already a new proto-professional body for property valuers which is looking to the West for models for its codes of practice and ethical standards. In Poland, land valuers are now being trained at the University of Oltzcyn, and a system of state licensing has been introduced.

But we certainly must not underestimate the cultural resistance to such developments which are now and will continue to be seen by many as reactionary: the bourgeois small landowner, the kulak, is still a bogeyman today; a real middle class, whether property owner, entrepreneur or professional adviser, may be much more than either the politicians or the people will stomach for some considerable time. The painful experiences look set to continue for many years to come.

References

Andonov, G. & Rizov, M., 1999. Land Reform and Land Market in Bulgaria, in: Dixon-Gough, R.W. (Ed.), *Land Reform and Sustainable Development*. Ashgate.

Arnison, C.J., 1984. *Leasehold Tenure as a Mechanism for Moral Governance*, in: *Land Management: New Directions*, Spon, London.

Braverman, A., Brooks, K.M., & Csaki, C. (Eds) 1993. *The Agricultural Transition in Central and Eastern Europe and the Former USSR*, World Bank, Washington DC.

Brooks, K.M. & Lerman, Z., 1994. *Land Reform and Farm Restructuring in Russia*, World Bank, Washington DC.

Csaki, C., 1994. *Where is Agriculture Heading in Central and Eastern Europe?* Presidential Address to the XXII International Congress of Agricultural Economists, Harare, Zimbabwe.

EBRD (European Bank for Reconstruction and Development), 1995. *Transition Report 1995: Investment and Enterprise Development*, London.

Grossman, M.R. & Brussard, W. (Eds) 1992. *Agrarian Land Law in the Western World*, CAB International, Wallingford, U.K.

Munro-Faure, P.W., 1994. *Surveying and the Property Profession in Central and Eastern Europe,* unpublished paper presented to Conference on Developing European Professionals, University of Leeds, U.K.

Rydval, J. & Pesl, I. 1999. The Privatisation and Restitution Process in the Czech Republic, in: Dixon-Gough, R.W. (Ed.), *Land Reform and Sustainable Development.* Ashgate.

Thiesenhusen, W.C., 1994. *Landed Property in Capitalist and Socialist Countries: Nature of the Transition in the Russian Case*, Land Tenure Center, University of Wisconsin-Madison, USA.

Annex 1: Excerpts from 'Transition Report 1995', EBRD, November 1995

Progress to Market Economies: the view of the European Bank of Reconstruction and Development,

Albania: 'Laws are drafted by legally trained personnel. Private parties generally believe that courts would recognise and enforce their legal rights against other private parties (but) they do not believe that courts would enforce such rights against state parties. For property that has been privatised, the law prescribes co-ownership between the new and former owners.'

Armenia: 'Laws are drafted by legally trained personnel. Independent legal assistance is available but from a very limited number of lawyers. Private parties generally believe that courts would not recognise and enforce their rights against state parties. Development of new private enterprises has been hampered by incomplete legal framework..'

Azerbaijan: 'Laws are drafted by legally trained personnel. Little legal assistance is available to private sector clients. Private parties generally believe that courts would not recognise and enforce their legal rights against state parties. Little progress (in privatisation) has been made.'

Belarus: 'Private parties reputedly believe that courts would not always recognise and enforce their legal rights against state parties and are sometimes reluctant to protect their interests judicially.'

Bulgaria: 'Few land titles have been issued, but 60 per cent of agricultural land has been handed back to the original owners through 'final land decisions' recognised as ownership documents and accepted as collateral. Private parties reputedly believe that courts would recognise and enforce their legal rights, including against state parties (but) rules restrict foreclosure procedures..'

Czech Republic: 'Laws are drafted by legally trained personnel. Sophisticated legal assistance is available, at least in Prague. Private parties generally believe that courts will recognise and enforce their legal rights, including against the state. About 30,000 industrial and administrative buildings, forest and agricultural plots and 70,000 commercial and residential units have been handed back to the original owners.'

Estonia: 'Privatisation of large and small enterprises has been virtually completed. Privatisation of Housing.. has been slow mainly because of time-consuming land surveys and legal uncertainties.'

Georgia: 'Private parties generally believe that courts would not recognise and enforce their legal rights, whether against a state party or another private party.'

10 The privatisation and restitution process in the Czech Republic

J. RYDVAL and I. PESL

Introduction

The former socialist Central and East European countries are all undergoing fundamental economic and political changes. Socialist regimes of these countries, based politically on the leading role of the Communist Party, economically on the state ownership as a means of production, and on a centrally planned allocation of resources, have collapsed. The political changes in late 1989 and throughout 1990, have created the conditions for the transition of the centrally planned economy to the market economy throughout the entire Central and East European region. The resulting complicated and structured process of economic reform is proceeding in three main spheres; institutional, monetary, and real. The implementation of this transition is determined by a series of steps carried out by the state. The system reform generally involves:

1. denationalisation (de-etatisation) and privatisation of state ownership;
2. price liberalisation;
3. foreign trade liberalisation and foreign investment;
4. tax reform;
5. reform of finance and currency relations (inclusive of internal and, later, full currency convertibility).

The basic reform steps are mutually conditioned and essential, but the sequencing of the steps, the methods, and the intensity of their application may differ between countries.

The change in ownership of the means of production is the basic condition for the rise of a market economy. Appropriate procedures for reconstruction of property rights by means of privatisation, contributes towards the activation of the public in the field of enterprising and ownership engagement in business activities leading to the evolution of the competitive market. The new owners become directly interested in the effective management of their businesses and their profit.

Denationalisation and privatisation is the most important and complicated step in the process of system reform and is significantly influenced by the quality of the cadastre. This is why this topic must be considered in detail. In the first part of this paper, the necessary legal framework for restitutions and privatisation of real property will be discussed together with the extent of the process. The second part will be devoted to the Czech land cadastre. This will document the changes of the cadastre in connection with the demands of the market economy. The third part will consider land consolidation and valuation of real property.

The privatisation process

The historical background

The process of denationalisation and privatisation of state property in the former Czechoslovakia started very early after the 'velvet' revolution, in November 1989. This programme was approved by most political parties in the elections that followed in 1990 and have been widely supported by politicians and the public up to this present day.

As in other former socialist countries, the economy of Czechoslovakia was almost entirely in the hands of the state. It can be said that the extent of nationalisation was extreme. For example, in 1989 only 1.2% of the labour force, 4.1% of the GDP, and 2% of all registered assets were in the private sector.

The privatisation strategy of the Czech Government is based on the recognition that the privatisation of a large part of state property is a necessary precondition of a transfer to the market economy. The speed of the transformation of ownership clearly affects the speed of the reform of the whole system. It is for this reason that the government wishes to realise the transfer of state property in the shortest possible period of time and its choice of privatisation methods, therefore, reflects this goal. Privatisation in the Czech Republic is a combination of restitution, small-scale privatisation, and large-scale privatisation.

Restitution

Restitution, i.e. returning the nationalised property to its former owners, has two main forms:

1.   the natural restitution of property;
2.   the financial restitution.

When possible, the form of natural restitution is preferred. In cases where the nature of the property has been significantly altered since nationalisation, the claimant may opt for financial compensation or else receive the property and pay, or be paid for the difference between the official estimated price of the property before and after the alteration.

Several laws having several common features regulate the restitution of the property belonging to naturalised persons (i.e. citizens). The common features are:

1.   only property unlawfully dispossessed after the communist take-over of 1948 is to be returned;
2.   the restituent (claimant) is a naturalised person (not a corporation or association) and a Czechoslovak citizen with permanent residence in Czechoslovakia (since 1993, a Czech citizen with permanent residence in the Czech Republic);
3.   the property is given back to original owners, heirs and family members;
4.   the restituent has to submit his written restitution claim to the present possessor in a delimited term specified in each law. These terms are preclusive. The claimant concludes with the possessor consequent agreement, that has to be approved by an authority (Land Office, former State Notary or Cadastral Office). The present possessor (in most cases a state owned enterprise or a co-operative) has to hand over the property within a specified period of time.

The most important of the laws containing these features are:

1.   **Law No. 403/1990 on Mitigation of Certain Property-Related Injustices**, which was the first law on privatisation adopted by the Czechoslovak Parliament in 1990 and which solved, above all, the restitution of the private apartment buildings nationalised during the last wave of nationalisation at the end of the 1950s.

2.   **Law No. 87/1991 on Out-of-Court Rehabilitation,** which covered a wide range of wrong doings suffered by Czechoslovak citizens under the communist regime. In the sphere of property rights, it repealed the unlawful administrative acts of the state authorities. This law partly restituted smaller industrial enterprises and other property, with the exception of property involved in the Land Law. The government has estimated that to the end of 1992, about 30,000 properties (mostly smaller enterprises) have been returned to private enterprise from nationalisation that was carried out between 1948 and 1955. In addition, approximately 70,000 items (mostly apartment buildings) have been returned to private ownership from the property that was nationalised between 1955 and 1959.

3.   **Law No. 229/1991 regulating the Restitution of Land and Other Agricultural Property (called Land Law)**. The previous owners are in accordance with the law, having given back agriculture land, forests, houses and agricultural buildings, and other agricultural possessions. The Land Law regulates restitution of the following categories of property:

    a.   dispossessed by land reform;
    b.   confiscated in connection with criminal prosecution for political reasons;
    c.   given up as forced sales and donations;
    d.   refused as inheritance under duress;
    e.   given up to ownership of farming co-operatives according to the Law relating to the Farming Co-operative System;
    f.   confiscated after illegally leaving the country;
    g.   in further enumerated cases.

The Land Offices must approve agreements on conveying the property by decisions.

A majority of the farming land and forests remained in private hands even after the socialisation of agriculture but the private plots, amalgamated into large land complexes, were cultivated and exploited by the Co-operative Farms, State Farms, and State Forest Enterprises without paying rent to their owners. The owners were divested of the right to dispose of these plots.

The Land Law stipulates that, unless a different agreement is reached between the legal claimant and the user of the property, a legally binding rental relationship arises on the day of the restitution of the property. This clause gives the current user the right to rent the land at officially set prices. The owner now has the right to cancel this rental agreement and to dispose of the land alone.

Administration of the property, and especially of the land that remained according to the Land Law in the state ownership, secures the Land Fund. The Fund administrates, assigns and leases the property until the privatisation.

Changes to the ownership and leasehold of the land are colossal and deeply affect the activities of the cadastre. In the period up to the end of September, 1994, the Land Offices filed 933,556 restitution claims in the whole Czech Republic, of which 115,463 cases were completely settled and in 179,697 cases, the Land Offices have already made decisions. Restitutions of state-owned forests and agricultural land covers approximately 33% of the total territory of the Czech Republic (some 24,000 sq. km).

The deadline for submitting the restitution claims of naturalised persons has already passed. Only the claims of Czech emigrants, who are Czech citizens, are currently being resolved. However, the extent of these claims is very small. The Central Land Office estimates the full end of the restitution process in approximately the year 2000 and the end of the main wave of cases by the end of 1995.

The laws that restore the property of legal persons have another philosophy. Institutions are generally excluded from the purview of the restitution programme. Restitution of the property of legal persons was carried out only in a few extreme cases, e.g. to sports organisations, communities and to some extent, to the Catholic and Jewish Church. The remainder of the former property of the Church is blocked to other forms of privatisation and the

Parliament solves the volume of the property that will be restituted

4.   **The Law No. 172/1991 on the Transfer of Some Objects from State Property to the Ownership of Municipalities** solves the problems relating to the property of municipalities. Municipalities were constituted as legal persons by the Law No. 367/1990 on Municipalities, and by the Law No. 172,1991 they were given back their historical property. In addition, they were granted the ownership of the state property they had administered until the validity of the law. Municipalities now have a large volume of real property, and Cadastral Offices have to register it within the cadastre relating to the municipalities within a short time.

Etatisation

As well as restitutions, the state also performed some cases of reparation by etatisation, i.e. the nationalisation of the property. In this process, the property formerly owned by the Communist Party and by the Socialist Youth Federation, acquired unlawfully from the state budget during the communist regime was nationalised.

Privatisation

The state property that could not be restituted was largely privatised by means of the so-called process of small-scale and large-scale privatisation. The simpler process involving the privatisation of the smaller units came first followed by the more difficult process of privatising medium sized and large state enterprises.

The process of small-scale privatisation was launched by the **Law No. 427/1990 on the Transfers of State Ownership of Certain Things to Other Legal or Naturalised Persons (Law on Small-Scale privatisation)** at the end of 1990. The purpose of the small-scale privatisation was the transfer of property forming compact business units. This included property in the service sector, production, the retail trade and local craft (with the exception of agricultural production) to the ownership or leasehold of Czechoslovak citizens (naturalised persons) or legal persons, the participants being naturalised persons only. These properties were not subject to restitution. The business units were privatised under this law in public auctions. District Privatisation Committees nominated by the Ministry of Privatisation conducted auctions. The small-scale privatisation process practically ended in 1993, adding 38 billion Crowns to the state budget and creating 22,000 new business units.

The largest volume of state property was privatised in the process of large-scale privatisation and was regulated by several laws. The most important law is the federal **Law No. 92/1991 on Conditions for Transferring State Property to Other Persons (Law on Large Scale Privatisation)**. This law covers the privatisation of state-owned enterprises such as banks, insurance companies and other similar financial institutions together with shares and other property assets owned by the state in other enterprises. The law regulates conditions and the process of the transition of state property to the ownership of both domestic and foreign naturalised and legal persons, for the purpose of entrepreneurship.

Large-scale privatisation is carried out either through several separate privatisation methods or through a combination of them. They include the:

1. transformation of state-owned enterprises into private companies, especially joint-stock companies, and a transfer of shares, e.g. through voucher privatisation;
2. direct sale to a predetermined buyer;
3. public auction;
4. public and non-public tender;
5. free (of charge) transfer to municipal ownership;
6. free transfer for the purposes of social security and health insurance.

The privatisation of each enterprise is carried out according to an officially accepted privatisation project. After the privatisation project has been approved, usually by the Ministry of Privatisation, the enterprise is transferred to the Fund of National Property. The Fund uses methods proposed in the approved project to sell or transfer the ownership of the property. Assets and yields of the Fund are not the income of the state budget.

Czechoslovakia was the first state that applied, to a mass extent, the method of voucher privatisation. In the Czech Republic 1,865 state-owned enterprises, with total book value 340 billion Crowns, were privatised in two waves during 1992 and 1994, respectively. 1,000 investment points were available to every interested Czech citizen in each wave, which were then exchanged for the shares of selected, privatised enterprises. The bidding for the shares of enterprises was organised in several rounds. Before each round, the authorities announced the share prices for each enterprise in terms of investment points together with the number of available shares to be offered in the next round. The price was set in such a way for each round so that it corresponded with the likely demand. If demand exceeded supply, no transaction took place and the investment points were returned to their owners. Voucher privatisation brought Czech citizens, free of charge, about one third of the property approved by the state for privatisation, within a short time period.

The extent of privatisation

The structure of property approved for privatisation up to November 30, 1994 (Hospodarske Noviny, 1994 No 250, p. 6) and privatisation methods are listed in Table 10.1.

With this relationship between private sector and the GDP, the Czech Republic leads the other Central and East European countries.

Transformation

For completeness, it is also necessary to carry out the transformation of agricultural, production, consumption and housing co-operatives to ensure that the residues of their etatisation and socialisation is removed and that they are transformed on the basis of an enterprise economy. The process of transformation is regulated by the **Law No. 42/1992 on Regulation of Property Relations and Dealing with Property Claims in Co-operatives (Transformation Law)**. From the point of view of the cadastre, the transformation of agricultural and housing co-operatives is very important.

This law did not solve the problem of the transition of flats and apartments to the ownership of the members of housing co-operatives in a satisfactory manner, and therefore new regulations for this purpose were introduced by the **Law No. 72/1994 on the Ownership of Flats**. This law also sets conditions for recording the ownership of flats in the Land Cadastre. The major problem lies in relating the different ownerships, and changes to the ownership, to a defined plot and to houses and blocks of flats that are located within those plots.

Table 10.1 Property and privatisation methods

Methods	Volume In million crowns	Percentage
Auction	9.079	1
Public tender	28,577	3
Direct transfer	92,633	10
Joint-stock company	745,820	81
Free transfer	49,147	5
Total value	**925,256**	**100**

Source: Hospodarske Noviny, 1994, 250, p. 6

The privatisation process and the Land Cadastre

New requirements

It was not only the restoration of law and order but also the economy that was the principle and the most urgent reason for implementing a governmental programme of massive privatisation and restitution. One of the basic conditions for a successful programme of privatisation and restitution of real estate property, was the existence of a reliable working land cadastre and land registry. Virtually all the authorities, institutions, companies, municipalities and citizens were asking for documents proving their ownership, not only of present properties but also historical ones for restitution purposes. The existing land cadastre proved to be quite insufficient and administration and work of the cadastral offices nearly collapsed as a result of the extra work load that this entailed. The ownership information required existed but the search process required highly skilled, manual and time-consuming work. It was clear that the process of cadastral reform was inevitable and that a modern cadastre is one of the basic condition for progress in the governmental privatisation and restitution programme.

The cadastral reform

The first and very difficult step in the process of cadastral reform was in the sphere of legislation. During 1991 and 1999, new cadastral legislation was prepared and came into effect from the beginning of 1993. Thus, January 1st, 1993 is a milestone in the long history of the Land Cadastre of the Czech Republic. The new basic cadastral legislation consists of the following laws and regulations:

1. Law No 964/1999 (essentially some changes and amendments to the Civic Code and other relevant laws);
2. Land Registration Act No 265/1992;
3. Cadastral Law No 344/1992;
4. Survey and Cadastral Administration Act No 359/199s;
5. Cadastral Regulations No 126/1993.

The basic cadastral legislation was completed in 1994 with the new Surveying Act No 200/1994. The new legislation creates the "legal cadastre", comprising both the former Land Cadastre and Land Registry, and uniting technical and legal aspects into one tool administered solely by the surveying authorities. After more than 40 years, the administration of the land cadastre has returned to the well established, basic principles of the previous Land Cadastre and Land Registry, although in a modernised form enabling the computerisation of records.

The real estate cadastre of the Czech Republic contains information about cadastral units, parcels of land, buildings connected to the land by solid foundations, flats and other rooms (although only in written documents), the owners and others in rights and their addresses, and finally the legal relations to the real estate property. The geometric determination and determination of the of the real estates is given in co-ordinates in the cadastre, either by measured data from field sketches, or by the representation of the property in the cadastral maps.

126

Cadastral documentation is arranged within the cadastral units in the following files:

1.   geodetic information file (cadastral maps and their digital data, if any);
2.   descriptive information file (written documents);
3.   summary surveys of the land fund;
4.   survey documentation;
5.   collection of deeds.

The rights to the real estate property (ownership, mortgages, encumbrances and other rights of material character) are registered in the cadastre as titles on the basis of deeds submitted to the cadastral office. There are three types of registration:

1.   entry (for all the agreement-based changes);
2.   record (for all the decisions made by state authorities and courts);
3.   annotation (for some informative facts).

Acquisition or conveyance of the real estate property on the basis of agreement does not come into affect until the property deed has been entered into the cadastre. Entry and deletion from the cadastre is in accordance with the Public Administration Act and is only possible on the basis of the decision made by cadastral authorities after the deed has been examined. The interested parties have a right to appeal to the Court against any negative decision. The registration in the cadastre has legal consequences and is considered to be correct unless the opposite is proved. Cadastral documents are open to the public and anyone has a right to examine them and to make copies for their personal use. Copies from the cadastre, made by the cadastral authorities are still to be considered as public deeds.

In accordance with the Survey and Cadastral Administration Act, the supreme administration body, submitting directly to the Czech Government, is the Czech Office for Surveying, Mapping and Cadastre. On a regional level, 7 Survey and Cadastral Inspectorates were created, whilst on the district level, 76 Cadastral Offices were instituted as the executive authorities subordinated to the Czech Office for Surveying, Mapping and Cadastre. In addition, there are 2 special institutes; the Surveying and Mapping Office, and the Research Institute of Geodesy, Topography and

127

Cartography.

The total number of employees has been substantially raised during the last few years and in 1995 there were 5,917 employees (5,385 working in Cadastral Offices). The key task for future is to introduce new employees into cadastre and to train all employees, step by step, for new technologies involved. The main role for the present generation of experienced and skilled staff employed in the Cadastral Offices, is to complete the basic computerisation of the cadastral documents and in leading the training programmes for new employees.

All activities related to the development and performance of the cadastre are funded from the state budget, since cadastral authorities are a part of the state administration. The fees collected for deed registration and for supplying extracts and verified copies from the cadastre, as well as for purchasing cadastral data (computer files or whole map sheets) are considered as the income of the state budget. The massive investment that is needed in all aspects of the cadastre is significantly supported by the government since it is clear that a reliable, working cadastre is one of the basic conditions for the restoration of the economy. Since 1990, the budget has increased nearly five-fold.

The new legislation and relatively good finance conditions are only the beginning of the long-term process of re-building the cadastre. The contents of the cadastre have to be completed according to the new legislation and all cadastral data (initially, all the descriptive information) computerised and in an easily accessible manner. The automation of the previous land cadastre does not cover the most relevant data, legal information about rights are still (in the most cases) submitted as hand written forms in the "owners folios".

A specific problem of the new cadastre is parcels of individual owners that had previously been amalgamated, according to past laws into huge agriculture or forest blocks. Such parcels are neither represented in current cadastral maps or in written documents, and are arranged only in the form of auxiliary files requiring searches for them in previous cadastral maps and documents (under the old parcel number). According to the cadastral law, such parcels should be entered into the cadastre not later than when the incident land consolidation is finished and, with respect to automation and the reliability of the cadastre, it is necessary to complete them as soon as possible. Since, however, there are approximately 8 million missing parcels in this category, the first step should be to complete the auxiliary files and integrate them with written cadastral documents, as a temporary solution to this long-term problem.

During 1994, the last cadastral offices were equipped with PC LAN (40 PCs on average, for more than 100 work places) and the supply of efficient work-stations for computer graphics commenced. Simultaneously, with the supply of hardware, development work on the cadastral software of a new generation started and transfer format standards were fixed.

In the same year, the long-term cadastral development programme was adopted and its implementation commenced in all cadastral offices. The key tasks of the programme are as follows:

1.   completion of missing parcels;
2.   conversion of owners folios;
3.   completion of tax information to parcel;
4.   densification of minor geodetic control (according to the requirements of land consolidation);
5.   digitisation of cadastral maps.

Progress in this field is likely to be limited by the considerable amount of current cadastral work, and this can only be overcome by substantially increasing the number of staff in the cadastral offices.

Cadastral documents for restitution and privatisation

The cadastre is the alpha and omega of the restitution and privatisation process. In the introductory stages of the process the input cadastral information is needed, whilst as the process develops and is completed, the results are registered in the cadastre. At the beginning of the restitution and privatisation process large quantities of cadastral information were issued to the public, not only from the current cadastre but also from the cadastral archives and from the former Land Registry. This initial stage is now nearly completed. Since the resulting restitution and privatisation deeds for registration in the cadastre are delivered at a rather slower speed, the present stage of consolidation should be fully used for the conversion and completion of cadastral data.

The extent of privatisation and restitution in the Czech Republic has no parallel in its known history. The same is also valid for the requirements of the cadastre. However, in spite of mans problems the present cadastre is still standing up to this crucial test.

Land consolidation

The historical background

Czechoslovakia underwent full collectivisation of agriculture based on the Soviet model. The inheritance of socialism resulted in the formation of large, inefficient co-operatives and state farms with high costs of production, an unsuitable infrastructure and large complexes of land. These did not respect the demands of nature protection and caused serious degradation of both the land and water quality. Present changes of ownership and leasehold of land, combined with the changing structure of agricultural enterprises resulting from the division of co-operatives and state farms, have given rise to family farms and the growth in the number of small users. This has, in turn, led to changes in production procedures and nature protection that has required the development of a new organisation for the land fund and the whole countryside. This is the main justification for initiating the process of land consolidation, which has a rather different character from similar procedures in countries with broken tenure. Owing to the transformation of agricultural enterprises, restitution and privatisation, large complexes of land are split into smaller parts

The legal framework for land consolidation

The above mentioned Land Law No 229/1991 deals with land reform and land consolidation on a common level. The process of land consolidation is characterised in this instance as a change in the arrangement of land plots or parcels in certain areas, designed to create suitable economic units for the land-owners in accordance with the demands of country creation, nature protection and capital construction. The District Land Office makes decisions relating to the process of land consolidation and on changes of owners' rights only after the agreement of those owners affected by the scheme. If it is necessary, the Land Office may allot an owner plots or parcels of other owners, on a temporary lease. The participants of the land consolidation scheme are able to appeal against a decision of the Land Office in the Court.

The process of land consolidation and the establishment and function of the Land Offices is regulated by the Law No 284/1991 on Land Consolidation and Land Offices. From the point of view of the cadastre, this law is very important. Land consolidation is defined here as an arrangement of ownership and easement (servitude) of parcels, a spatial and functional arrangement of the parcels through their amalgamation, subdivision, re-allotment, boundary alignment and ensuring access to all parcels. In addition, conditions for rational farming, for the protection and fertilisation of the land fund, for improving the land and increasing its ecological stability are created simultaneously. Land consolidation is, whenever possible, implemented throughout the whole cadastral unit.

Two forms of land consolidation

There are two forms of land consolidation:

1.   simple land consolidation;
2.   complex land consolidation.

the first being actually a land subdivision (the first step), and the second a land re-allotment (the second step as the result of the entire process).

The process of simple land consolidation is of a temporary nature and is intended to ensure that land is returned for private farming purposes as rapidly as possible, particularly when the parcels or plots are very small. The owner is, in effect, given back his own plots. The shape of these plots is often long, very narrow and not very suitable for farming. When it is not possible to give the owner back his own land, the Land Office allots him a similar area of plots of other owners on a temporary lease. This is free of charge (the plots normally having been part of a co-operative). In areas where larger plots of land have been allotted to land owners', a process of consolidation is applied, the boundaries being aligned and accessibility of parcels improved.

Complex land consolidation is carried out in the cadastral units where a substantial part of the possession of land was returned to the private farmers. This process can only commence when the owners of more than half of area of the agricultural land within the scheme agree to its implementation. This type of consolidation includes creating suitable economic units of land, the construction of joint facilities, technical, hydro-economic, fertilising and ecological measures. Complex land consolidation is the main tool of the program of restoration of the rural parts of the country.

Land consolidation commenced in 1991. At present, the process of simple land consolidation operates in most cadastral units. Approximately 22,000 schemes have been started and about half of them have been settled, in most cases by the allotment of parcels on temporary, free of charge leases. During the third quarter of 1994, the process of complex land consolidation was commenced in 264 cadastral units. Most were concentrated in districts of the region of South Morovia where the soil is very fertile soil and the population very interested in private farming. The introduction of further complex land consolidation schemes is largely dependant on the availability of finance from the state budget. The Central Land Office considers that an adequate approach would be to take 2 cadastral units in every district each year.

The District Land Offices open and manage the process of land consolidation. They solve the problems of reclamations and, after the closing the oral dealing, legalise in writing the decision relating to the results of land consolidation.

The participants of the process of land consolidation are:

1.   the land-owners, whose parcels are included in land consolidation;
2.   other naturalised and legal persons, whose rights can be touched.

The land-owners elect a body of representatives that co-operates on both the proposal and the realisation of the land consolidation. A proposal of land consolidation is ordered by the Land Office and carried out by a naturalised or legal person with permission to project land consolidation. The proprieties of the proposal are regulated by the Regulations No 427/1991 on Land Consolidation.

In carrying out land consolidation, an exchange of ownership rights to parcels is often necessary. Criteria for this are outlined in the Regulations. For example, it is not permissible to exceed a 3% difference in price and a difference in area of 10% among exchanged parcels without the agreement of the owner. The difference in distance between the original and exchanged parcels is also taken into account. It should not exceed 20%.

Land consolidation and the cadastre

The Land Office ensures within the realisation of the proposal of land consolidation the following factors:

1. the restoration of old boundaries and the setting-out of new boundaries;
2. the maps or plans of the reallocation survey.

These are submitted to the Cadastral Office as documents for updating the cadastral database. The cost of the full implementation of the proposal of land consolidation, including the surveying work, building of access roads and the protection and restoration of the country, is borne by the state. The remaining cost related to the land consolidation is borne by the land-owners with possible state subsidy.

When the process of privatisation began, the advice of the surveying and cadastral experts were largely ignored. In spite of many reminders, the adopted laws almost led to the collapse of the former Centres of Surveying (now re-named Cadastral Offices) in that they were not able to provide information in reasonable terms and to update the cadastre. The participants realised that close co-operation was the only solution. Co-operation between the branches of surveying and cadastre, represented by the Czech Office for Surveying and Cadastre, together with the Ministry of Agriculture and the systematic management of the Land Offices, has resulted in significant improvements. This has resulted in the Agreements on Co-operation of the Cadastral and Land Offices, in 1993, that solves the problem of co-operation in the restitution of property and land consolidation. The Provisional Instruction for Surveying and Projection Work in Land Consolidation was introduced in 1994 to solve problems relating to the technical and organisational aspects.

In the simple land consolidation, there are of two ways in which changes may be made to boundaries. The first method involves the preliminary staking-out of boundaries which are classified as temporary changes and are, thus, not subject to the cadastre. The second method involves permanent changes that are marked in the cadastre according to standard subdivision plans, in which all the original parcels, and their parts within the new parcels, are represented. This latter method is more suited to small fragmented parcels and cannot be used without difficulty in whole cadastral units in the case of complex land consolidation, and it is not even necessary.

Complex land consolidation has to be carried out in accordance with the Provisional Instruction. The first step in this process of land consolidation will be the determination of the perimeter or boundary of the consolidation project. The input data for the consolidation process involves the characteristics of the plots of the single owners which are balanced by taking into consideration the factors outlined above, i.e., area, distance, price before consolidation. The equivalence of the exchange of plots is considered by comparing the total balance of plots of the same owner after land consolidation (output data). It is possible to use the same method in both a simple land consolidation and on a larger scale. Once the process of land consolidation has been completed, new digital cadastral maps and descriptive data files are created. It is presumed that in the remainder of the cadastral unit (excluding urban areas) the digital cadastral map will be consequently completed.

This process is complicated by the huge volume of work and the present state of the cadastre that imposes limitations upon the possibilities of using modern surveying instruments and computerisation in surveying world and in the design of land consolidation. A rapid computerisation and digitisation of the cadastre may eliminate some of the defects of the previous real estate cadastre. This is one of the reasons why a long-term programme of cadastral reform has been adopted.

Valuation of real estate property

Liberalisation of the economy has become evident in the real estate market as well as in productive farming land. After more than forty years without any free market, there are many problems and misunderstandings related to the valuation of real estate which are, according to the Civic Code, the land and buildings connected with the soil by solid foundations

Since 1939, the price of real estate has been strongly regulated. Even agreed prices were under state control and had to be determined according to the price regulation then in force, under the penalty of absolute invalidity of the deed. For nearly 20 years, the task of the valuation of real estate has been reduced to the mechanical application of the price regulations, i.e. to simple, routine calculation of the administrative price. As

134

the price of land was frozen and only the prices of buildings were allowed to develop, the prices of buildings reached, as a rule, much higher values than the proper land. Thus, the assessment of real estate became the domain of expert witnesses mostly from civil engineers rather than surveyors.

Due to the long period of price control, the contact with the reality of the free market was entirely lost and these, very simplified administrative prices had nothing in common with actual free market prices. After the liberalisation of the economy and the elimination of price regulation, a very complicated and peculiar situation arose. An unknown and changing economic environment influenced, to a great extent, by 'grey economics', and the inability to compare actual market prices, are the main reasons for the present unsteady, free market price levels of real estate. In addition, a huge amount of real estate was sold during privatisation for distorted administrative prices or, in many cases, even lower prices. The initial extreme increases in the prices of real estate, no matter where situated, stopped and the prices started to vary in a very significant but random way. The process of consolidation that is needed will take many years and it is pre-conditioned by opening to foreign capital and by releasing a price of leasehold. The conditions of such a pseudo-market do not support the general understanding of economics and the transition to a free market. A considerable amount of insufficiently covered loans and mortgages given by prestige financial institutions, without any knowledge of the actual or possible values in the free market, has resulted in heavy losses, delaying tactics, and demands of exorbitant guarantees.

Generally, several types of pricing policies and values are used and are requested in the valuation of real estate, depending upon the function of the property.

An administrative price is used as the basis for assessment of notary fees, of conveyancing, inheritance and gift fees, and for other purposes defined in connection with restitutions, privatisation, land consolidation and for some compensations.

An agreed price is applied to the market, and its valuation is needed for real estate agencies, finance institutions for guarantees of loans and mortgages. Some banks are requesting, in addition to the current price, a valuation of the instant liquidity price.

For tax purposes (real estate tax) the simplified valuation on the basis of a tax return, according to special law and regulations, is used. The tax assessment is not derived from the actual price and can be considered as a form of administrative value.

The present administrative price is according to the Price Law No 526/1990 and the Regulations No 178/1994 about Valuation of Buildings, Land and Permanent Vegetation. This applies some principles and techniques in order to get closer to the actual market prices, involving e.g. price maps in urban areas and soil quality maps in rural areas. Due to the long-standing absence of a free market environment, a more reliable valuatlon of real estate still remains a too distant goal.

Administrative prices according to the previous regulations (now generally ineffective) are demanded for some special cases of compensation in the processes of restitutions and land consolidation.

Conclusion

The extent of the restitutions and privatisation has no parallel in the whole known history of the country. Enormous political pressures for fast progress of the whole process have been managed to support the development of the cadastre, which has proven to be essential to its success. On the other hand, the cadastre has been exposed to many political and other pressures from the very beginning of the whole process. There were both tendencies to simplify relatively complicated cadastral procedures at the expense of credibility and reliability of the cadastre, and even to break the law. The only way to stand these enormous pressures is to adhere to the legislation and to the basic technical principles absolutely essential for sustainable maintenance of the cadastre. During the last two years, the Cadastral Offices have been given the unusual authority and confidence of the public. But only the future can verify the results.

While the restitution and privatisation process has been practically finished in general outline, except the agricultural sector, land consolidation is needed to solve the agricultural land completely and represents a long-term process for decades.

Valuation of real estate in the environment of the free market is at the very beginning. Its development is closely conditioned on the progress and liberalisation of the national economy.

References.

Cihelková, E. & Fingerland, J. 1992. *Komparativni ekonomika: Stredni a Východni Evropa (Comparative economics: Central and Eastern Europe).* University of Economics. Prague. 125pp. *In Czech.*

CÚZK, 1994. Koncepce digitalizace katastru nemovitosti (The conception of digitisation of the Cadastre). *Zpravodaj CÚZK,* **2**: 2-6. *In Czech.*

Kaulich, K. 1994. Kprubehu restitucía pozemkových úprav v roce 1993 (Restitution process and land consolidation in 1993). *Zemedelské Noviny,* **25**: 6. *In Czech.*

Kluson V. 1991. *Remarks concerning privatisation.* Institute of Economics of the Czechoslovakian Acadeny of Science. 81pp.

Kotrba, J. & Svejnar, J. 1993. Rapid and multifaceted privatisation: experience of the Czech and Slovak Republics. *CERGE Working Paper Series.* Charles University, Prague. 58pp.

Liska, P. 1994. *Príruka pro vlastníky pudy a jiných nemovitosti (A handbook for the owners of land and other real estate property).* Prospektrum, Prague. 398pp. *In Czech.*

Pesl, I. 1992. Ekonomie, právo a evidence nemovitosti (Economy, law and the Land Cadastre). *Econom,* **XXXVI(20)**: 56-57 and **XXXVI(21)**: 52-53. *In Czech.*

Pesl, I. 1994a. The re-establishment of the Land Cadastre in the Czech Republic. Presented paper at the UN Workshop: *Data Processing in Cadastre,* Vienna.

Pesl, I. 1994b. Remarks on the quality of data in the legal Land Cadastre. Presented paper at the UN Workshop: *Data Processing in Cadastre,* Vienna.

11 Conceptions of rural planning following land privatisation in Hungary

J. NYIRI and R.W. DIXON-GOUGH

Introduction

With the change in the political system, which occurred in 1989/1990, there arose the possibility of a significant change concerning the use and ownership of land, leading to a major programme of land privatisation. Many national and international institutions were convinced that the role of private ownership was one of the most important factors in the growth of the economy. For the transfer of land from the state to the private sector to be accomplished successfully, a reliable infrastructure for the registration of ownership had to be established (van Hemert, 1993). The principle considerations related to the land policies developed during this period are:

1. acknowledgement by the government that the private land/real estate market is an essential requirement for the working and development of the economy;
2. within a reasonable period, legislation and executive mechanisms for the working of the land/real estate market should be working reasonably well across the whole country;
3. the organisations involved in land registration should, where possible have a level of private enterprise;
4. the costs of the land registration process should be funded by the users and, in general, the organisation should be self-financing.

In Hungary, with its reputation of 'ersatz communism', the basic organisational structure was developed in the early 1970s. However, due to a low priority of ownership and changes of ownership at that time, only land and property data were computerised. In the latter part of the 1980s, land reform and the moves towards a market-orientated economy increased the level of interest in the development of a country-wide, computerised solution for land registration and cadastral mapping. Land registration in Hungary is administered by the Ministry of Agriculture. The area of Hungary is approximately 93,000 square kilometres and, at present, there are about 60,000 cadastral map sheets at a variety of age, scales and projections. Each separate land parcel has a unique number and this is used to link the registration records to the cadastral maps. There are currently about 6.5 million parcels, but this number is being steadily reduced by a process of land compensation and consolidation. The programme of land reform being undertaken in Hungary is similar to that of other former Communist Block countries. Land that was formerly owned by the State is being sub-divided and used to compensate former landowners and other claimants.

Various solutions for land privatisation (such as compensation, the arrangement of proportionate property, and the division of parcels between former employees and members of co-operatives) have resulted in the former, large agricultural holdings being divided into many small parts and, in some instances (because of the simple reason that the parcels are rarely of a convenient size and shape) becoming useless for economic production. The situation demanded a full and detailed review of the tasks of land consolidation.

This chapter will present examples of the current status of the established property structure in Hungary, present the legal and financial background to land consolidation, consider the background to the TAKAROS project and finally, present the TAMA land consolidation project. These tasks present a series of actions that pose a number of questions related to agrarian policy, the financial economy, land evaluation, judicial and technical tasks, and land registration. Despite the rapid progress that has taken place since 1989, solutions are still being sought for these questions and this is particularly true in the case of determining the future of the villages of the agrarian communities.

The establishment of the current land property structure since 1945

Within this chapter, it is impossible to cover the entire technical and legal development of the Hungarian land property structure since 1945. It will, however, identify and indicate the most significant developments that have taken place during this period.

The devastation caused during the Second World War was not restricted to the urban areas but affected rural areas also, and agriculture in particular. Nevertheless, in 1945 moves were made to modify the traditional land structure and divisions in an attempt to eliminate feudal inequality. Hungary was the first nation to achieve this. The enforcement of **Law 600/1945** was the largest project of land consolidation in the history of Hungary, and this process touched virtually every community. In total, 34.8% of cultivated land was parcelled and distributed to 650,000 people, the average parcel size being 5 Hungarian acres (2.875 ha). Despite the scale and extent of the process it was, however, impossible to satisfy all legal requests. Part of this process of agrarian reform included the areas of domanial and state-owned forests and these were used for the establishment of state forestry farms. This took place from 1948. From this time, the centuries' old dream of Hungarian peasantry had been realised in the provision of the small land parcels. One of the problems that resulted from this form of distribution was, in effect, that the parcel sizes were too small for cultivation, except for vegetable crops.

This problem was overcome in 1949 by the organisation of co-operatives, in which the small land parcels were combined - often against the wishes of their new owners. This process continued with the formation of more and larger co-operatives. After 1956, changes in the political structure lead to an increased demand for the creation of larger co-operatives, and eventually a co-operative was established in every settlement. The political changes of this period laid an unacceptable burden and inconvenience upon the rural community. Agriculture had now, once again, become 'large estates', with huge agricultural concerns being formed by the fusion of smaller units. One co-operative, in particular, eventually had a holding of 15,000 ha.. In addition to political pressures, this process of concentration of the agricultural holdings was enforced by **Law 4/1967**.

During the 1970s, the law on land consolidation was modified (**Law 22/1976**) and the basic organisational structure, currently in place, established. However, due to a low priority of ownership and changes in ownership at the time, only the land and property data were computerised. As part of this law, the subject of land consolidation widened to include:

1. general (when it related to the entire area of a settlement);

2. partial (when only a part of a settlement was included in the planning process);
3. between collective farms;
4. within collective farms.

The law of 1976 introduced a new category of land that not used by collective farms for large-scale production. This was to be allocated to local people and was to be termed 'permanent land use'. In this context, it is significant to note that in 1995, it was estimated that 22% of Hungary's national wealth was in the land. In the past, agriculture was the supporting pillar of the people's economy and it must be so in the future.

Further legislation was introduced in 1987 with **Law 10/1987**, which regulated changes to land ownership and **Law 11/1987**, which regulated land consolidation. The areas of land consolidation legislated for in the 1976 with the introduction of **Law (22/1976)**, which identified the general conditions of land consolidation based on similar principles. Tables 11.1 and 11.2 illustrate the distribution of land in Hungary according to property and land use, respectively.

In the latter part of the 1980s land reform, and the moves towards a market-orientated economy, increased the level of interest in the development of a countrywide, computerised solution for land registration and cadastral mapping. The Ministry of Agriculture administers Land registration in Hungary. The area of Hungary is approximately 93,000 square kilometres and, at present, there are about 60,000 cadastral map sheets at a variety of age, scales and projections. Each separate land parcel has a unique number and this is used to link the registration records to the cadastral maps. There are currently about 6.5 million parcels, but a process of land compensation and consolidation is steadily reducing this number. The programme of land reform being undertaken in Hungary is similar to that of other former Communist Block countries. Land that was formerly owned by the State is being sub-divided and used to compensate former landowners and other claimants.

During 1989 and 1990, significant political changes took place in Hungary, as they did in all former socialist countries. One of the political parties, the Hungarian Democratic Forum, established an agricultural programme during the summer of 1989. This addressed the important step of changing the ownership of the land.

The Hungarian Parliament passed the first Land Compensation Act in 1991 (**Law XXV/1991**), which was further entitled, "The Law on Land Ownership Changes in Order to Compensate the Damages that were Unjustly Caused in the Citizens Ownership by the State". At present, the work of the compensation is slowly coming to a close.

Table 11.1 The distribution of land in Hungary according to the property form, in May 1987

Property Form	Hectares	Percentage of available land
State Property	2,658,044	28.57
Land use by co-operatives	5,690,145	61.15
State Farms	279,238	3.00
Complementary Farms	539,783	5.80
Private Property	114,187	1.23
Other Property	21,779	0.25
Total	**9,303,176**	**100.00**

In 1990, the Institute of Geodesy, Cartography and Remote Sensing initiated a plan for the development of 115 District Land Offices (DLOs) and 19 County Land Offices (CLOs). This was further developed by the Department of Lands and Mapping of the Ministry of Agriculture and presented at the 1991 FIG Conference. It was accepted by the EC for PHARE funding (Remetry-Fülöpp, 1996). The project involved the computerisation of the land register including office automation, introduction of uniform application data handling, and the loading of data relating to about 6.5 million parcels, and the computerisation of cadastral mapping data management at DLO level with integrated geodetic measurement and control point data.

In 1994, a further land law (**Law LV/1994**) was passed, which regulated the ability of the individual to change the land property of one's own free will, to purchase land or property, to change the use and utilisation of the land, and to protect the land.

Table 11.2 The distribution of land in Hungary according to agricultural land use, in May 1987

Land Use	Hectares	Percentage
Arable land	4,709,323	50.61
Gardens	338,335	3.64
Orchards	96,524	1.04
Vineyards	144,861	1.56
Cereal crops	1,222,239	13.14
Reedbeds	40,046	0.43
Uncultivated land	1,056,455	11.36
Fishponds	26,427	0.28
Forest	1,668,966	17.94
Total	9,303,176	100.00

The present state of the land structures after the Compensation Act

Under the regulations of the first Land Compensation Act of 1991 (**Law XXV/1991**), persons who had previously experienced losses were given compensation vouchers to an amount proportional to their losses. The people were not given back their original lands but could use the vouchers to bid, or 'buy back' land from the co-operatives or state farms. A process of re-distribution through auctions therefore, formed the basis for their new land holdings.

The value of each compensation voucher was 1,000 Hungarian Forints (HUF), which corresponded to 1 Hungarian Gold Crown (this being an equivalent value of the land based upon a number of characteristics, such as fertility and location). Thus, the value of 1 ha of first-class land might be

143

valued at 40 Gold Crowns. Within the first Land Compensation Act, three types of land holdings were identified; those for the Land Compensation Act (i.e. those people who had either worked at, or were employed in a co-operative), for Proportionate Owners (those people who were working in a co-operative who were entitled to a share from the properties of the co-operative), and a further set of holdings for members of co-operatives (the employees of the co-operative or state farms - who are given 20-30 Gold Crowns worth of land). The size of the land holdings, together with the number of owners, is listed below in Table 11.3.

From Table 11.1, it may be seen that under the first Land Compensation Act, the land holdings for compensation (from agricultural co-operatives and state forestry farms) has totalled some 2.5 million hectares, whilst 5.25 million hectares had been privatised by 1995. Thus, of the whole of Hungary available for cultivation (approximately 8.7 million hectares), more than half had been privatised.

The ability of the owner to cultivate the land had no bearing upon either the acquisition of the land and the size of the land holding. It was entirely dependent upon the title of buying and the value of the potential owner's compensation vouchers. Examples are given below of four Hungarian counties (Békés, Baranya, Pest and Somogy), each having entirely different characteristics relating to position, situation and agriculture (Figure 11.1). In each county, the desirability of land would appeal to different buyers and for different purposes. Békés is a largely agricultural county, where the main products are maize and wheat. Such an area would have little appeal to those not directly concerned with agriculture. The county of Pest is near to the capital Budapest and here, the attraction is largely recreational. People bidding for land in this area would typically be interested in small plots for garden and weekend allotments - for the cultivation of fruit, vegetables and vines. Lake Balaton is located in the county of Somogy and this is a tourist resort and wine production area, in which much of the land is purchased with a view to building holiday homes. Finally, the county of Baranya is important for both industry and agriculture.

Table 11.3 The land holdings and ownership of land in Hungary (as at 1995)

Ownership	Land Holdings (millions of hectares)	Number of owners
Compensation	2.45	526,700
Proportionate owners	2.10	1,760,000
Employee owners	0.50	380,000
Subsequent compensation	0.20	100,000
Total	5.25	2,766,700

Figure 11.2 is an example of the present state of the Land Compensation Act. This particular example is of the village of Nagynyárád in the county of Baranya. It may be seen from the example that many of the land parcels are very long and thin (typically 6.5m wide and 570m long) and there are some extreme examples in which the width of the parcel is 0.25m. A similar situation can be found in any other country. In each county, bids were made, by auction, for the land parcels and the value of the bids for 1 Gold Crown ranged from 500HUF to 2,000,000HUF (on the shores of Lake Balaton).

The situation regarding the bids for the land may be summarised below as:

1. many large fields had to be sub-divided into smaller parcels before they could be bid for;
2. the size of many of the land parcels around the big cities and holiday resorts were less than 5 hectares;
3. around the outskirts of many villages, the parcels were even smaller, between 1 and 2 hectares in size;
4. it is not possible to farm economically on such small parcels, even when the parcels form part of a small to medium size farm;
5. many of the small parcels, forming part of a farm (or the holdings of an

individual) are located in different places - either around a village or even in different villages;

6. market competitiveness is affected by both the relative size and distribution of the parcels and the expense of producing crops is influenced by this land-property structure. Furthermore, the expense of farming within this structure influences the competitiveness of agriculture on both domestic and foreign markets. The small parcels are unable to produce marketable goods unless their expenses are reduced to a minimum.

Figure 11.1 The location of the counties within Hungary

The Land Office of Hungary plays a major part in the privatisation of land and the compensation of former owners. Hungary's Compensation Act ran out in 1995, and this placed much pressure on the system of land registration, particularly in Budapest where there was a backlog of 280,000 applications in 1994 (Alverson, 1994). Five million hectares of land have been affected by land reform. All changes should be surveyed, registered and updated in the land register.

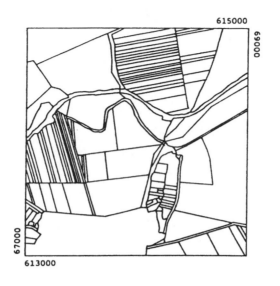

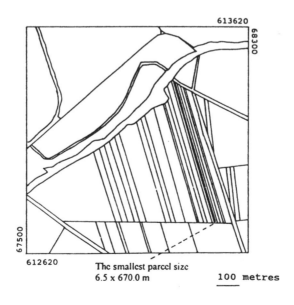

The smallest parcel size
6.5 x 670.0 m

100 metres

**Figure 11.2 The current state of land parcels in Nagynyárád, Baranya
after the Compensation Acts**

To cope with the rate of such changes, a multi-purpose land and property register and digital cadastre is mandatory for an effective transition to a market-driven economy. Such a system has been developed and is called TAKAROS.

TAKAROS

TAKAROS is an acronym for a Hungarian phrase meaning, "map-based cadastre system for computerisation of land offices". This is a new project that is being developed, partly in recognition of the wider commercial possibilities of base mapping and land registration but also to address the problems of the creation of over 2 million new land parcels. In parallel to TAKAROS, the land register in the Capital Land Office (CLO) has also commenced a programme of computerisation. In parallel to TAKAROS, the land register in the CLO has also initiated a programme of computerisation. A pilot cadastral map renewal project in two districts of Budapest was launched, with the help of bilateral Swiss funding and expertise, in 1995 (Papp, 1996). Subsequent developments to TAKAROS will include parcel-based regional Land Information Systems (LISs) with Geographical Information System (GIS) capabilities, inter-operability between all Land Offices and remote access for municipalities, notaries and lawyers.

The present system is based upon a related, yet separate cadastral and land registration system. This is maintained through the network of 115 DLOs. In these, the legal title to property is recorded on a property sheet under three headings, the description and the location of the parcel, the history of the ownership of that parcel, and any other legal facts, such as rights and restrictions. The legal boundaries of the properties are defined on a cadastral map, each parcel being uniquely numbered to provide the link between the map and the property sheet.

This system is being superseded by the TAKAROS project. Ultimately, this project will provide a nationwide LIS, both at county and district level, the exception being the city of Budapest. Here, a separate initiative has been started by the city's land office to address the special problems encountered in the city.

Eventually, the Budapest and countrywide solutions will be harmonised. TAKAROS has been designed to cover the functions of land registration, applications for registration, property sheet management, map management, cadastral map management, geodetic control point management, survey record management, and GIS services and other derived products. These various functions have been separated into those to be administered by the DLOs (land registration and map management), and those to be administered by the CLOs (GIS and other activities).

TAKAROS was designed to provide an integrated LIS, thereby allowing the district offices to integrate the land registration records within a maintained, topologically structured parcel fabric. All spatial data is stored in Oracle to ensure that the relationship between the parcel fabric and the property sheet is always consistent. Any operation relating to parcels are performed by interactively manipulating the parcel structure yet preserving the integrity of the dataset until the changes have been approved. This is aided by the use of a single database, thereby allowing reliable data-sharing and relatively easy maintenance and administration. Particular attention has been paid to the long-term value of the data to ensure that it will be compatible with future technologies with little conversion needed.

For such a system to operate smoothly and effectively, a defined workflow has to be followed. Every document received by the district offices is subject to a unified application registration system to ensure a standard entry point. This has three functions; to direct the accepted application to an appropriate TAKAROS sub-system, to check the result of the workflow, and to notify the client about legal decisions and possible problems. Within this, a property sheet management system allows the property sheets to be modified, where necessary, whilst the geodetic and control subsystem (GEO) maintains the geodetic and control point database.

All survey and cartographic data for TAKAROS is processed by the survey and map data entry sub-system (SUR). This also converts data from earlier base data (on different geometric frameworks) to the unified national projection system. A common database links this to other subsystems, particularly GEO and to the cadastral map management subsystem (MAP), which is responsible for the display of the digital base-map. MAP is able to download data from the central database into local graphic files. Any map data input into TAKAROS enters the digital database via SUR, and since all parcel records are maintained in a strict sequential order, the history of the individual parcels, together with the entire parcel fabric, is preserved.

With any system that allows data entry at different levels, a safety net is required. TAKAROS, never deletes data: data when superseded is retired and stored in the database. Normally, only the active data can be viewed but under certain circumstances, it might be desirable for a DLO to display the historical picture of the parcel fabric together with the appropriate property sheets. During this operation, is it not possible to retrospectively modify the data. In districts where some parcels are being extensively and repeatedly modified or sold, the volume of data might be too great to store in the database. In these cases, archival data is offloaded to external devices.

TAKAROS is, at present, only producing the software management for digital cadastral map management. The ultimate aim of the National Cadastre Programme is to provide digital data for all land offices and to store this in the TAKAROS database. This will require an extensive re-survey of all areas where existing cadastral maps are not of an acceptable quality. TAKAROS is just the first step towards the full computerisation of Hungarian Land Offices and its digital records must be accessible and reusable for a long time. The system has been designed to enable multi-purpose use of cadastral and land registration data. It is planned that in future TAKAROS will have:

1. linkages to external data bases, particularly to census and socio-economic databases;
2. further development of its wide area network (WAN) for CLOs and DLOs;
3. WAN capability among all land offices.

Eventually, it is intended that TAKAROS will provide data access to the whole country form any land office, enabling land registration procedures and queries to be made locally, without having to travel to the district within which the property lies.

Both GIS and LIS require data for operation, yet the level of funding for cadastral mapping in Hungary was and is relatively low. To overcome this problem, the National Cadastral Programme was launched to seek external financial resources. Setting up an independent, self-financing government agency could provide a solution for an integrated approach to Land Information Management (LIM). To assist in this, a German Federal Bank together with the Bavarian state government has provided a loan to promote the renewal of Hungarian cadastral maps, 50% of which are based on old projection systems. The top priority will be given to settlements in the GIS National programme, larger cities, town and the villages affected by the TAMA land consolidation pilot projects (TAMA being an acronym of, in Hungarian, land consolidation in Hungary). The National Cadastral Project comprises the acceleration of the data entry to the land registers, digital renewal of the large-scale base map, continuation of GIS-supported land consolidation, monitoring land use, completion and renewal of the 1:10,000 topographic map base, and assessment of new land valuation techniques.

A computerised Land Office service, based on the approach of an integrated LIM, will make a direct contribution towards the strengthening of the property market by ensuring security of title, improving legal authenticity, supporting credit and facilitating transactions. It will also meet user requirements, especially those of the local governments and utility companies. Some of the areas in which the Ministry of Agriculture is a potential client are: land-related statistics, land consolidation, land use monitoring, data provision for soil conservation, vegetation mapping and crop-related analysis, agricultural supervision and control, forest management, landscape and regional development with an emphasis on the delineation of less favoured areas.

The present state of the redistribution of land-property

At present, the two opportunities that are available in Hungary for changes to the land or property are defined according to the Land Law of 1994 (**Law LV/1994**), which regulates the process of land consolidation. Only with the free will of the farmers can changes be made to the structures of the land or property. Applications may be made to the government for the process of land consolidation form either private landowners, contractors and the presidents of co-operatives. In 1994, for example, 15 applications were made - the majority coming from the county of Hajd-Bihar. The aims for land consolidation ranged from the founding of private and family farms, reasons based upon special agro-economical grounds, to the 'creation' of collective farms.

Once the process of land consolidation has been agreed upon by all the individuals concerned, a civil code can be issued and the ownership of the land can be changed by contracts of sale and purchase. It is important to bear in mind that there is no fixed time-scale for land consolidation. It is a large task,which has to be carried out in compliance with the programmes of regional agrarian development (Fenyõ, 1997)

The future for land consolidation

Land consolidation must be initiated by the landowners and this step can only be taken at the consent of all the owners involved in the process. Only the collective landowners have the right to initiate this process. Reasons include the desire by the land owners to consolidate a number of small holdings into a single large parcel, to be able to reach all parts of the consolidated parcels easily and conveniently, whilst increasing the economic value of the holding. To achieve this, the landowners need financial and expert support and assistance.

Alternatively, a claim may be made by a commune or village to amend the general plan of the settlement, again with the consent of all individuals and families involved in the process. Typically, this could involve redesigning the road and track networks of the village, altering and regulating the field irrigation system, determining the minimum acceptable size of parcel, and prescribing the agricultural land use.

It is evident from enquiries relating to the consolidation of land and property that changes are needed and the only real solution will be to launch a massive programme of land consolidation that can only be resolved through a voluntary process of exchange. Similarly, the tasks can only be performed through regulations although the legal environment for this process remains to be ascertained. Once the land holdings have been effectively consolidated, further property exchanges or sales may be made, thus increasing the effective production of agricultural goods and land. The advantage of land consolidation may be found, not only with the private sector, but also with the state. More efficient farming methods will produce more products, which will generate more revenue, leading to greater amounts of tax being collected, thereby improving government income.

The determining factors for land consolidation are:

1. an increase in the number of properties and holdings. All auctions for the realisation of the Land Compensation Acts, should be completed by the end of 1998. Surveying works, concerning the creation and setting out of the new parcels, should be completed by 1999. Only when the parcels have been surveyed can the data relating to them be recorded in the Land

Register, and this is expected to take several years. At present, there are more than 6 million parcels and the compensation process will create a further 600,000 new parcels. From co-operative titles (proportionate ownership) a further 700,000 parcels will be allocated to new owners, many of whom will then lease their parcels to others. If the task of land consolidation is to proceed in an organised form, then the parcels and their ownership should be registered without delay;

2. the compensation of the land offices (the installation of computers and software) is now complete. This should speed up the processes of land registration, which until the completion of computerisation had been overwhelmed by the work originating from the Land Compensation Acts and privatisation. In addition, the land offices are now under great pressure to maintain the demand for security of ownership;

3. demands for accurate multipurpose cadastre maps in a digital form will need to be increased because of the growing needs and requirements of the new owners.

To satisfy the points outlined above, new legislation is required for land registration; the existing legislation consisting only of a decree with legal force, which was modified to permit the use of computers in 1995. At present, both surveying and mapping are also regulated by an executive decree and this too must be replaced by legislation. A new Hungarian Land Law was passed in 1995, which permits a limited degree of voluntary exchange of land parcels. This still, however, requires some minor modifications. At present, agricultural land has no real value and landowners cannot use the land as security for loans. Legislation needs to be introduced to allow mortgages and credits to be raised on the value of the land. This will not only aid agricultural production, but will improve the economy of the nation. Figure 11.3 illustrates the concepts, processes and stages that should be ideally involved in a new Land Consolidation Law.

The TAMA land consolidation project

Some of the concepts illustrated in Figure 11.3 are currently being tested and evaluated as part of the TAMA project (TAMA being an abbreviation, in Hungarian, of Land Consolidation in Hungary). An agreement has been signed between Germany and Hungary to develop TAMA under the EU's

PHARE project. The main partners in the project are, from the German side - the Bureau of Soil Evaluation in Kiel and, from the Hungarian side, the Land Offices from the counties of Békés, Baranya, Pest and Somogy.

These particular regions were chosen on the basis of their agrarian economies. The experts form both countries have analysed the current situation in the respective countries on the basis of the characteristics of the agricultural products, the natural environment, the availability of existing data.

The processes involved in the development of the TAMA project are illustrated in Figure 11.4. The realisation of this land consolidation project necessitates the development of a system of digital land registration and a full coverage of digital cadastral maps.

Digital land registration

The first module of the complex, decentralised land registration system has already been installed in the 116 District Land Offices and the installation of the database is well under way. Once the system is fully installed and functional, it will be continually modified and extended. The system will allow a wide range of users (public notaries, financial institutions, local government, private surveyors, etc.) to access the system through remote terminals.

Digital cadastral maps

Demands for digital maps are steadily growing, both from local government and from the various agencies involved in the compensation process. Digital map production will be speeded up following the passing of legislation concerning the national cadastre programme. Initially, the task for the National Surveying and Mapping Agency will be the primary production of the layers for the "Technical Framework Map" (the base map). Once this has been completed other scales of digital maps will be produced - with particular attention being given to cadastre maps. In some cases, several map scales might be simultaneously produced for an area. Figure 11.5 illustrates the process of land consolidation, based upon the digital land registry and the digital cadastral map, within one of the counties involved in the TAMA project.

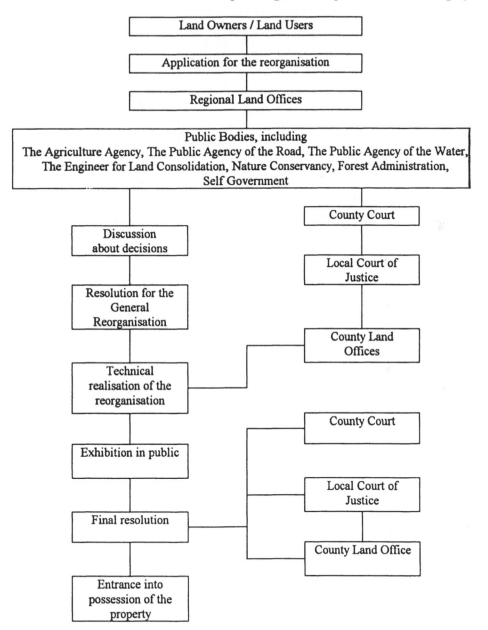

Figure 11.3 The processes involved in the Land Consolidation Law

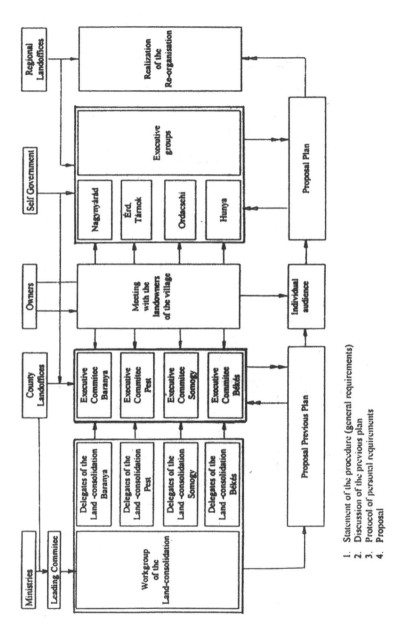

Figure 11.4 An outline of the processes of the TAMA project

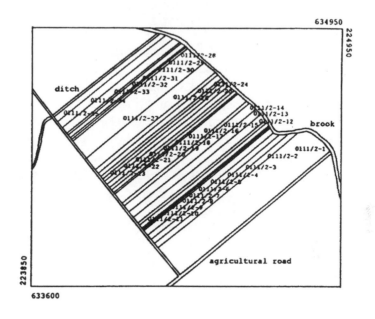

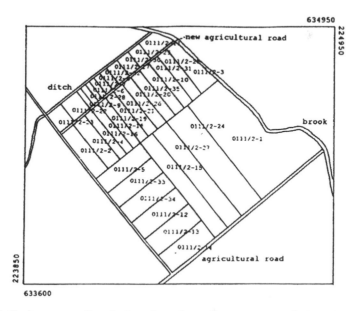

Figure 11.5 An example of the planning of a new parcel structure based on digital data

Conclusion

Because of the problems encountered in Hungary, of a state emerging from a long period of communist domination, there are many tasks concerning the reorganisation of land, which must be addressed. The ultimate goal is to establish a complex land consolidation programme that functions in a similar manner to many other European countries. Owing to Hungary's limited financial resources, only one segment of the TAMA project - the reorganisation of landed properties can be realised in the first phase.

There is a great need for legislation incorporating the new concepts concerning the reorganisation of landed properties.

References

Alverson, C. 1994. Knocking on the door of Europe and the GIS mainstream, *GIS Europe*, 3(9), 22-23.

Bogart, Z. 1990. *Mat keel Tudni az Agrárprogramról*, MDZ Füzetek, MDF Országos Elnöksége, Budapest.

Fenyõ, G. 1997. State of land ownership and land utilisation in Hungary. Presented paper at the Technical Seminar in *National Land Tenure Development in Eastern and Central Europe*, Bertinoro, Italy.

van Hemert, J. 1993. Land registration in eastern Europe and the fall of communism, *Geodetical Info Magazine*, 7(9), 75-79.

Hoffer, I. 1985. Negyven esztendõ, *Geodézia és Kartográfia*, 2, Budapest.

Joó, I. & Szabó, Gy. 1989. Podgatovka kodrov v oblasti zemleizerenija i zemleustrojtv v Vengerskoj Norodnoj Respublike, I, *Seminar Socialistikih Stran po Voprosom Obuchenija Studentov i Issledovanijam v Oblasti Zeleustrojstva*, Olsztyn.

Joó, I. & Nyiri, J. 1995. Present state and efforts of the Hungarian agricultural policy including the role of the College for Surveying and Land Management. *23rd International Symposium of the European Faculty of Land Use and Development*, Olsztyn.

Laws, 1991. évi XXV. Törvény a tulajdonviszonyok rendezése érdekében, az állam által az állampolgárok tulajdonágtalanul okozott károk részleges kárpótlásáról.

Laws, 1994. évi LV. Törvény a Termõföldrõl.

Nagy, S.I., 1979. A földrendezés sajátosságai és igazgatási modellje, *Geodézia és Kartográfia*, I, Budapest.

Papp, I. 1996. Taking stock with Takaros, *GIS Europe*, 5(10), 36-38.

Remetey-Fülöpp, G. 1996. Towards integrated land registration: renewal of the Hungarian cadastral system, *International Journal for Geomatics*, 10(6), 32-35.

Sepsey, T. 1990. *A Kárpótlási Törvényről*, MDZ Füzetek, MDF Országos Elnöksége, Budapest.

Szabó, Gy., 1983. Föld- és területredezés I-II, *EFE Földmérési és Földrendezõi Fõiskolai Kar*, Székesfehérvár.

Szabó, Gy. & Ágfalvi, M. 1988. *Flurbereinigung in Ungarn, Flurbereinigung in Europa, Europaischen Fachtagung Flurbereinigung*, Landwirtsschaftsverlag GmbH, Münster-Heltrup, 361-400.

Váhl, M. 1949. Áttenkintés az 1945, évi földreformró és annak végrehajtáról. *Az állami földmérés Közleményei*, **I**, évf.

12 Land tenure and land reform in South Dobrudja (north east Bulgaria)

G.ANDONOV, M. RIZOV and K. BATANOV

Introduction

The pattern of land tenure and ownership in Bulgaria has been complicated by the country's political history. This is particularly true in the case of South Dobrudja. Within the complex web of inter-related phenomena that affect or are affected by current developments, this paper specifically considers the important factors in the history of land tenure, the current situation of land reform, together with their implications on one of the most important agricultural areas of Bulgaria. Whereas the conditions in Dobrudja are slightly different to the general picture of the country, there are a number of useful conclusions that might be drawn from the experiences of this region.

The beginning of modern land tenure systems

Dobrudja is situated in the north eastern part of Bulgaria and the south eastern part of Rumania. During the time of Turkish rule this region was an integrated territorial unit. The split came after the Russian-Turkish war of 1877-1878 when it was divided into two parts; the northern territory, which was left to Rumania, and the southern, which was included in the new Bulgarian State.

The area of Bulgarian Dobrudja is approximately 722,000 ha and has been divided into six administrative units (okolii); Balchic, Dobritch, Tervel, Silistra, Dulovo and Tutrakan. The establishment of the new, independent state of Bulgaria meant that capitalistic economic principles replaced the former Turkish feudalism. Immediately after this war, the agricultural sector was effectively ruined and the vast areas of land, abandoned by the Turkish ex-landlords, populated by poor people.

The most important landowner in the region became the State, which controlled about 100,000 ha of land - that of the emigrating Turks, Tartars and Cherkes - through the "Act of Tartars' and Cherkes' Land" of 1880. This was known as the State Land Fund. Approximately half of this land was sold off to the Bulgarian population, over the following thirty years. A further proportion of the land, comprising mainly pasture land with a limited amount of forest, came into the ownership of the municipalities. At the end of the nineteenth century, approximately 156,000 ha of land were owned by the municipalities. The remainder of the agricultural land was privately owned and consisted of many relatively small farms. The breakdown of private ownership is illustrated below in Table 12.1.

In the eastern part of the region almost half of the private arable land was owned by about 8% of the landlords, with an average farm size being in excess of 50 ha. In the western region, however, about 50% of the land is under the ownership of three quarters if the landlords, the average farm size being less than 10 ha. Just 800 persons control 125,000 ha. (30%) of the privately owned land. In 1879, the area of leased land consisted of 40,000 ha., including 9% of privately owned land and more than 90% of the State Land Fund and land owned by the municipalities.

This pattern of private ownership, however, changed quickly under a series of comprehensive internal and external factors. By the end of the nineteenth century, Dobrudja was characterised by small-scale farming, with about 50% of the farms being less than 10 ha, and a new capitalistic type of land tenure system.

Pre-war period 1900-1912

As a result of rapid and significant developments in trade, assisted by the convenient location of three ports located within the region (Balchik on the Black Sea, and Silistra and Tutrakan along the River Danube), the export of agricultural products – grain, cattle and poultry – proved to be very successful. This upward trend led to an increase in the number of land owners by some 48% and a corresponding increase in the area of arable land, by some 17%. As a result of these increases, by 1905 over a third of the large farms (over 400 ha) in Bulgaria were situated in South Dobrudja, whilst the overall amount of leased land had remained unchanged. The rent

161

on this was paid as a proportion of the farm-yield, between 30 and 50% of the yield, or in money-terms between 10 lv/ha and 26 lv/ha.

South Dobrudja under Rumanian rule

From 1913 to 1916 and from 1918 to 1940, South Dobrudja was invaded and occupied by Rumania. During this period a backward, semi-feudal land tenure system was introduced. In a law passed for the purpose of organising 'New Dobrudja', 30% of the privately owned land and almost all of the State Land Fund and municipality land was expropriated in favour of the Rumanian settlers. This particularly affected those land owners who had more than 100 ha of land. For those land owners with more than 1,000 ha, often 95% of their land was taken away, the only compensation being symbolic and paid in securities. In the whole, approximately 100,000 ha of land was expropriated from the Bulgarian population. About 10,000 Rumanian settlers were provided land, with an average farm size of between 8 and 10 ha, at the expense of the Bulgarians.

The overall number of land owners fell, during this period, by approximately 300 per annum and the number of people who were land-less rose to approximately 10,000. In 1940, data reveals that 32,596 land owners owned, in total 201,000 ha of the land. The greatest proportion of this was in farms of up to 5 ha (Table 12.2). The trend of leasing land was maintained since 48% of this land had been rented by farmers (mainly Bulgarians) owning more than 50 ha. The total amount of leased farmland had reached 93,600 ha by the end of 1940, and the majority of the leasing landlords were Rumanians living in the towns.

Owned land (in ha)	Eastern Part		Western Part		South Dobrudja as a whole	
	Landlords (%)	Land area (%)	Landlords (%)	Land area (%)	Landlords (%)	Land area (%)
up to 5	30.4	2.4	54.8	16.2	46.7	7.9
5 - 10	21.5	6.4	23.8	30.5	23.0	16.1
10 - 20	23.0	16.3	19.0	35.9	20.3	24.2
20 - 50	17.5	28.0	2.3	12.6	7.3	21.8
50 - 100	5.2	17.7	0.2	1.8	1.8	11.3
above 100	2.4	29.3	0.1	3.0	0.8	18.7
Total %	100.0	100.0	100.0	100.0	99.9	100.0
Total ha	10,000	249,018	20,000	167,002	30,000	416,000

Table 12.1 The breakdown of private ownership in 1900 (Source: the State National Archives)

Owned land (ha)	Farms (%)			Land area (%)		
	1908	1940	Recapitulation	1908	1940	Recapitulation
up to 5	42.3	64.0	+ 21.7	10.5	31.2	+ 20.7
5 - 10	26.9	22.5	- 4.4	17.3	22.9	+ 5.6
10 - 20	21.1	9.4	- 11.6	25.1	17.7	- 7.4
20 - 50	7.9	3.1	- 4.8	23.5	12.6	- 10.9
more than 50	1.8	1.0	- 0.8	23.6	15.6	- 8.0
Total	100.0	100.0	+/- 0	100.0	100.0	+/- 0

Table 12.2 The breakdown of private ownership in South Dobrudja in 1940 (Source: the State National Archives)

The Kriaiova Treaty and land reform

As the result of the Kriaiova Treaty (7th September, 1940), South Dobrudja was given back to Bulgaria. This led to a new, significant change in the system of land tenure. A total of approximately 84,000 people were evacuated from Bulgarian Dodrudja and some 67,000 immigrants accepted from other parts of Bulgaria. An exchange of land was carried out between the Bulgarian and Rumanian landowners, since approximately 20,000 ha of land within Bulgarian Dobrudja was owned by Rumanians. The Rumanian landowners disposed of this land, according to Article V of the Treaty, by 1947.

The Bulgarian government faced a complicated situation, whereby it was not possible to pay compensation for the land owned by Bulgarian citizens in Rumanian territory. Of the 238,000 ha of State Land Fund made available for immigrants from North Dodrudja, some 148,000 ha was needed to recover expropriated properties. Of the remaining 90,000 ha, 55,000 ha was distributed amongst those peasants with no land leaving a balance of 35,000 ha. This remaining amount of land was far less than than the 152,000 ha of land required to compensate the Bulgarian citizens for their land left in Rumania.

The average size of land ownership units in the 1940-42 period was between 7 and 9 ha, although there were some farms of between 400 and 800 ha (Table 12.3). Compared to previous years, the leased land did not indicate any rise but, in fact, declined by 50%. In 1945, the recorded number of landowners was 9,493 who leased 41,000 ha of land including that from the State. The leased land was cultivated by 29,908 farmers.

The socialist land reform

After 1945, a socialist land tenure system was established in Bulgaria. The Law of Labour and Land Ownership (1945) expropriated all land with an area exceeding 30 ha expropriated by the State. Subsequently, all land owners who were not themselves farmers were limited to 5 ha. In this way, a State Land Fund was established in the Dobritch subregion (located in the eastern part of Dobrudja) consisting of 54,103 ha of arable land, of which 21,120 ha was distributed amongst landless or small farmers.

Municipality	Agricultural land area (ha)	Formating units		Landlords	
		number	average size (ha)	number	average size (ha)
Balchic	10,755.4	1,582	6.8	954	11.3
Tervel	10,365.1	3,393	3.0	1,631	6.4
Dobrich	17,568.3	5,067	3.5	3,895	4.5
For the subregion	38,688.8	10,042	3.9	6,480	6.0

Table 12.4 The current distribution of agricultural land in South Dobrudja
(Source: Ministry of Agriculture, Sofia)

Owned land (in ha)	Eastern Part		Western Part		South Dobrudja as a whole	
	Landlords (%)	Land area (%)	Landlords (%)	Land area (%)	Landlords (%)	Land area (%)
up to 5	31.4	8.3	45.9	22.2	39.5	14.0
5 - 10	32.9	23.9	42.5	51.3	38.1	35.3
10 - 20	24.3	24.2	10.8	22.0	16.9	23.3
20 - 50	9.4	27.0	0.8	3.5	4.7	17.3
above 50	1.6	16.6	- 0.0	0.8	0.8	- 0.0
Total %	100.0	100.0	100.0	100.0	99.9	100.0
Total ha	25.503	258,470	31,781	181,789	38,284	440,259

Table 12.3 The average size of land ownership units in the 1940-42 period (Source: the State National Archives)

In South Dobrudja, during the period of 1945 and 1991, a number of large co-operatives were created. At the beginning of the period, these were operated independently in almost every town and village. However, after 1956, they were merged. In the Dobritch subregion, there were six co-operatives with a total area of 36,396 ha, together with six state farms having a total area of 38,612 ha. In 1970, as the result of the policy for achieving a larger scale of agriculture, the existing co-operatives and state farms were merged into eight agriculture and industrial complexes with a total area of 400, 120 ha.

The current situation of land reform

With the passing of the **Law of Ownership and the Use of Farming Land (1992)**, the process of restoration of land ownership rights to the original owners commenced. This law, by nature, is a restitutional one since it provides for the return of land ownership rights to the original owners as they were at the time of the creation of the co-operative farms, up to the amount of 30 ha per landowner.

In general, however, the lack of title proofs or too many of them of a contradictory nature have caused significant delays in the reform schedules across Bulgaria. In South Dobrudja, however, the process has been developing at a much faster rate compered to the remainder of the country. To date, from a total of 214 village territories, reallocation plans have been prepared for 209 of them and 202 are already legalised and under implementation. For example, in the Dobrich sub-region, the agricultural land has been distributed under a pattern illustrated in Table 12.4. These data present clearly the difference in land ownership structures between eastern and western parts of Dodrudja.

A specific feature of land reform in Dobrudja from 1991 has been the active management of state land. Each year, approximately 9,500 ha of land has been leased to tenants who cultivate 100 or more ha. Furthermore, about 4,000 ha has been made available to farm research institutes. During this period, two main trends can be identified; co-operative farming and the leasing of farm land.

Co-operative farming, to a large extent, replicates the traditional system of co-operatives, where all assets of the landowners, shares of the buildings, animals and machinery contributed. For this, a charge for rent is agreed at a meeting of various landowners. A disadvantage of this system is that, in practice, the members of the co-operative cannot exercise any control over the management process. Leasing of privately-owned land is a common feature in Dobrudja. This can be explained firstly, by the increased availability of the land of owners who are migrating from the villages to the towns (yet who wish to maintain the title to the land) and, secondly, by changes in the profession brought about by lack of finance, machinery and appropriate skills.

Private farms comprised of between 300 and 800 ha of rented land are becoming common. The term of lease is usually between 1 and 3 years with a rent paid, equivalent to between 20 and 25% of the net operational profit. In some municipalities, one can find farms comprised of up to 4,000 ha of leased land (in Gen. Toshevo) and even 20,000 ha in Krushari. A more generalised analysis, however, indicates a trend with the average private farm size, including rented land, being in the order of between 500 and 800 ha, and that of co-operative farms being between 800 and 1,000ha.

A current complication, according to LOUFL, is that the majority of restored land ownership rights are either to people who are too old to farm the land, or dead. Most of the rights are related to the original owners who pooled their lands in the co-operatives during communist times. This situation is further complicated by the lack of an operating land market in Dobrudja, despite some attempts at a few preliminary sales. The expectations are that in an open land market, the average size of owner-occupied farms will increase to between 20 and 30 ha, whilst other operating farms have an average sze of between 500 to 800 ha.

Conclusion

The distinguishing feature of the current land reform in Bulgaria has been the desire to execute the programme quickly. There is, at present, a general level of dissatisfaction amongst administrators who regard the rate of progress towards land restitution with some concern. Reasons for this can be sought in both the historic background of land tenure and the more efficient management approaches currently being adopted.

13 Land reform and the land market in Bulgaria

G. ANDONOV and M. RIZOV

Introduction

Land reform is a process directed towards establishing a free property market comprising not only the restoration (restitution) of land to the owners, but also the creation and development of a market economy and market mechanisms to direct the development of the country on a new way.

The realisation of the reform to be imposed has to be worked out and passed through a suitable legislation basis, in order to establish suitable government structures and units, together with financial and investment funds. On the other hand, the land reform is influenced partly· by the political climate, the actions of the government, the participation of international funds and investment.

Analysis of the status of land reform

The restoration of the ownership of farmland in Bulgaria was really prompted and legislated for by the **Law of Ownership and Use of Farmland (LOUFL)** passed in 1991. It was amended in 1992 and in 1995, which significantly changed a number of basic principles with regard to the restitution of land, its management and use. The numerous changes of the statutory and methodological basis not only delayed the process of land reform but also, in some territories and regions of the country, to a great extent fully blocked it.

Until now, the Municipal Land Commissions have recognised right of ownership in 5,379,400 ha of farmland. During the past 5 years many different kind and character of documents were found, of differing degrees of importance, relating to the processes of former land ownership. In many territories there are no existing documents at all, whilst in others there are only a very small number of documents. In some instances, the documents appear to be too contradictory and inconsistent to be recognised and accepted as being valid. Sometimes pressure has been exerted over the work of the specialists in the Municipal Land Commissions by the executive and political powers, and other interested bodies and institutions, such as Members of Parliament, etc.. In other cases, the incompetent interpretation of the law and subjective and unrealistic recommendations are increasingly confusing the contractors. This is further complicated in that the basis for recognising ownership in the different legislative acts serves as an additional obstacle and continues to impede the progress of land reform.

There are many other reasons for the delay in land reform, such as the unsatisfactory staff and technical equipment of the Municipal Land Commissions, the high number of modes of restitution, financial embarrassments and, not least, the still strongly influencing political partialities at the different levels of power in the country.

The natural climatic conditions, coupled with the different technical equipment of the contracting companies additionally serves to delay the implementation of the processes of land restitution. More recently, in the separate territories belonging to the settlements (TBSs), there is another phenomenon - the unwillingness of the owners to speed up the process of register the land quickly. This is connected with its use, fear from taxes and additional payments, inheritance problems etc..

The land reform in different types of regions in the country is at different stage of development and realisation:

Flatland farm regions with developed corn production (Dobrudzha, the Danube plain, Miziya, Trakiya, Zagore etc.) cover some 2547 TBSs (52.9% of the country) with 4,001,000 ha. (that is 74.4% of the land available for restitution). In these regions, the 1958 land reallocation plans were implemented for an area covering approximately 3,498,000 ha. comprising about 65% of the farm land in the country or 87.4% of the land in these regions. The situation in 1995 was:

1. land reallocation plans for 1418 TBSs covering 2,537,800 ha. (47.2% for the country or 63.4% for the flatland regions) have entered force;
2. 540 plans were subject to revision having a combined area 960,200 ha (17.8% of the country or 24% of the flatland regions);
3. 589 TBSs covering 503,200 ha. (9.4% of the country or 12.6% of

171

the flatland regions) are in the process of being work upon reflecting the conditions from the amended LOUFL;

4.   455,151 owners of 1,522,220 ha. (28.3% of the country or 38.0% of the flatland regions) were entering into possession during 1995.

Table 13.1 Land reform in Bulgaria in regions

Region	Number of TBSs	Prepared Plans		Approved Plans		TBSs with existing boundaries	
		No.	%	No.	%	No.	%
Burgas (SE)	482	320	66.4	234	48.6	182	37.8
Varna (NE)	535	414	77.4	350	65.4	169	31.6
Lovech (CN)	501	326	65.0	230	45.9	201	40.1
Montana (W)	400	243	60.8	123	30.8	203	50.8
Plovdiv (S)	496	232	46.8	143	28.8	252	50.8
Ruse (NE)	50	364	72.4	297	59.0	125	24.8
Sofia (SW)	913	371	40.6	178	19.5	699	76.6
Sofia town	68	17	25.0	11	16.2	34	50.1
Haskova (S)	919	353	38.4	205	22.3	501	54.5
Total	4817	2640		1771		2366	

Source: Ministry of Agriculture, Sofia (1995)

Mountainous and semi-mountainous regions - cover 47.1% of the country and has approximately 1,378,200 ha. of farmland within 2270 TBSs (25.6% of the land for restitution, which in the greater part consists of old, existing boundaries, easily restored and with small area of the plots, sloped and with significant areas unusable. In these regions, the situation at 1995 was that:

1. 672 plans with 547,000 ha. (10.9% of the farmland in the country) had already been announced;
2. plans to restore land within existing boundaries in 1068 TBSs with 794,500 ha (14.8% of the farm land in the country) had been entered into force;
3. 393 TBSs of some 174,500 ha. (3.2% of the country) were the subject to reworking;
4. plans for 879 TBSs covering 400,920 ha. (7.6% of the country) are in the process of being prepared;
5. 717,719 owners on 900,600 ha. of land (16.7% of the country) were entering into possession during 1995.

This information is summarised in Table 13.1.

It may be said, with some confidence that the status of land reform corresponds to the established and functioning legislation, organisation and political pre-conditions. The bulk of the land reform was expected to be completed during 1997, although that does not include some TBSs for which the process could be longer.

Conditions and pre-requisites for the progress of land reform

It was originally anticipated that the process of land reform in Bulgaria would be completed by 1995 but, unfortunately, this was not to be. The reasons for this were elaborated upon in the final report of Phase II of the EU's PHARE Technical Assistance Project. In general, the principal reasons for the delay were:

1. new realities and pre-requisites for the implementation of the reform for the restoration of ownership came about as a result of decisions made in the Constitutional Court, in 1995, concerning the amendments of the LOUFL;
2. the law did not determine either the optimum minimum and maximum parcel sizes and how they would be grouped to meet the requirements for modern agriculture. This, in turn, reflects the interests of society, particularly from the point of view of land management and land use;
3. the problems of land use were deliberately ignored and, in many

cases, the land owners were not able to implement their own choice of land use management;

4. when, in the majority of TBSs, the land is fragmented, it is not possible to apply modern agriculture practices to the small land parcels. It is, therefore, necessary to draft new laws for land consolidation, grouping and the protection of land - all of which take up considerable resources, time and personnel;

5. according to the new statistical data, between 25 and 30% of the land in the mountainous and near mountainous regions is either not farmed or farmed in a very primitive manner, compared to between 8 and 10% in the flatter, more fertile regions. Thus the land in the mountain regions cannot logically be co-ordinated with the other, more productive farming regions without modern machinery and an increase in productivity to allow prices of produce to be reduced to make them compatible to the European market;

6. during 1994 and 1995 there was a slight recovery in agricultural production, due mainly to more than 2,000 newly formed co-operatives and associations and between 1,500 and 2,000 family and individual farms. This recovery was, however, accompanied by many obstacles and difficulties. These included, problems with modernisation including obsolete and worn-out machinery, problems related to prices and the buying-out of farm produce, high interest rates for bank loans, and a lack of any support or guarantees from the State to revive this important sector of the country's economy, which has almost been destroyed. Livestock breeders and those farmers with perennial plantations were experiencing these particular problems. Valuable material assets have been lost and continue to be lost. These include vineyards, orchards, livestock farms, all of which are being destroyed.

An important contributing factor for this destruction is the inability of many Bulgarian land owners to separate the problems which are intrinsically connected and which cannot be considered separately, such as land, capital, machinery, livestock, processing, preservation and markets;

7. regardless of the declarations and announcements made in the press concerning the various aids from different foundations and programmes, neither land nor agrarian reform have received any tangible support or guarantees for future development.

It is logical to expect, after the restoration of land, that everyone would have the right to manage and use the land to their particular requirements. In reality, the land in Bulgaria will be formed by associations and co-operatives, with no particular regard to the unnecessary politicisation involved in the process. In Bulgarian villages, the young are departing to urban areas, leaving a growing, elderly population without either state support or the support of younger helpers to farm this valuable asset.

Although there are many problems concerning the development of land reform, progress is being made. Land restitution is a fact and the government's objectives are gradually being realised, together with the availability of resources, staff, restructuring, etc. The professionals working in this field are aware of the problems but their expectations and suppositions for improvement are optimistic, especially in relation to those processes related to landed property - registration, the property market, taxes and fees, investments, etc.

The land market

The normative acts that have been placed in force since 1990 give a real basis for establishing a free market of land with its different sectors - sales, leases, inheritance, conceding to landless, alienations etc. In practice, after the restoration of the ownership and receiving of the necessary documents, every owner can freely dispose of his land.

As a consequence of the existing provisions and in reality, there is a limited land market in the country. Despite the number of issued notarial acts (over 50,000) for the ownership of farm land, sales are currently at about 3% of ownership and not more than 1% of the total area of restituted land.

The majority of these are separate properties (Mostly having an area of less than 0.8ha) located by roads, highways, important industrial sites, resort areas, settlements along the Black Sea coast, mountain resorts etc., where the function of the land is being changed for the use of buildings, facilities, commercial areas, petrol stations, motels etc.

The restitution laws and the consecutive restitution of ownership is influencing, and will continue to influence, the land market bearing in mind the wish of the majority of those who became rich to invest their capital in immovable property. In this respect, the privatisation dealings attracts special interest, which to one extent or another, is depend on and influenced by the land market. This will possibly result in an additional fragmentation in smaller parcels, but initially this market is not directed to land with agricultural functions, but to the development of properties with more attractive potential uses, such as trade, tourism, recreation etc.

It is expected that a revival of the market will coincide with the approval and establishment of duty-free zones, zones with special relevance to the Balkan Region, the Black Sea zone etc.

In the context of the sales and purchases of agricultural land, this is essentially restricted to defined land use, mainly of farm buildings, livestock farms, machinery buildings and plots, auxiliary farms, and centres created the policy and the activity of the liquidation councils of the former collective farms.

In the farm regions, land is often leased for a periods of one to three years where, with the available material assets and machinery, the role of some owners is being reduced to that of lessees. On average, lessees rent from owners and from the state, farms of about 500 - 1,000 ha. In grain producing areas, the size of leased area increases to between 5,000 - 90,000 ha.. A more general analysis shows that an average farm based upon leased land ranges from 300 to 800 ha and only in a few cases to over 1000 ha.

In this respect, it is high time that a special law be established and implemented concerning the leasing of land and related tenure that will inevitably stimulate the land market, provide guarantees for both the owner and tenant, whilst defining the conditions of the market for agricultural commodities etc. It will also be necessary to change some provisions of the normative documents in effect referring to ownership, foreign investment, mortgaging, development, environment etc.

Conclusion

Land reform is passing through a number of different stages and is constantly changing and improving. There is a stabilisation of the conditions for restoration of ownership, easing the subdivisions among heirs, the possibility for the grouping and consolidation of land, the free market of properties and the conditions for development over farmland. Improvement is the normative basis for change of the function of farmland, the conditions for privatisation and the preferences for the owners and the users.

The market of land is a fact, despite the many imperfections, shortcomings and obstacles, and gives a real opportunity for dealings and development of the properties in the emerging market economy of Bulgaria.

14 Sustainability as a principle in spatial development? Questions for policy and education

F.B. ROSMAN

Introduction

In a relatively short time, sustainable development has become an issue in many areas. At the Delft University of Technology, a policy statement was adopted in 1994 in which the topic features prominently. As a result, it has become an evaluation criterium of the new curriculum which is currently being implemented. This leads to the direct question in which topics, having a connection with sustainable development, must be considered relevant to the students of the Faculty of Geodetic Engineering and the role these topics will play in their professional career.

The paper examines the issue of sustainable development in context of the curriculum of the Faculty, with special emphasis on the nature of Dutch land reform and spatial development, as relevant to the employment situation of geodetic engineers.

Sustainable development

Most sources agree that a description of the term sustainable development takes the general form of 'the art of living now, without harming the potential of future generations'. Although most of the long-term effects of production methods and emissions may not yet be provable, this does not mean that we do not have to act now. In essence, sustainable development is a choice, a factor in judgements may be made, and an ethical principle. It can only work properly if it is supported by public opinion and is implemented by those who design policies and measures.

In society, there are encouraging signs that this is the case, although these are principally connected either to strong convictions or the economic advantages of those involved.

Land reform and spatial development

In the Dutch situation, land reform can be described as the procedures necessary to change the land-use situation for the present owner, or land-user, to facilitate the implementation of public policies in the field of spatial development. This includes both urban areas and infrastructure.

These changes include buying, leasing or expropriating land and require appropriate compensations. It must be noted that in the Netherlands, (local) government often takes the initiative for new urban projects. This is certainly the case for major projects, which are in, or connected to, the larger cities. Although many possibilities for these types of land reform exist in law, stalemate situations can occur where private or institutional owners are interested in developing projects. It is also difficult to prevent land-speculation or attempts to privately develop a project, due to the open and democratic nature of the decision-making process.

In rural areas the initiative can come from several sources; groups of land-users, (private) nature preservation societies, and local, provincial or national authorities.

Sustainable development as a principle

Recent history indicates that the use of resources is largely dependent on technology and choice, with sufficient feedback to control the use of the resources. From the industrial revolution until very recently, such feedback loops have either not been present or were not sufficiently effective to act as such, and this has resulted in the over-exploitation of resources.

Choosing sustainable development as a principle of controlling the use of resources is absolutely necessary, but such a choice creates new problems in its application to small scale, short-term decisions. The lack of knowledge of appropriate (sustainable) levels of many environmental parameters makes it next to impossible to give a solid foundation to many decisions. Consequently, for both policy-makers and the business community, the temptation to do what is popular or profitable may be stronger than the (long-term) benefits to the earth's ecosystem.

The Governing Board of the Delft University of Technology have adopted a policy statement, a new strategic view, called *Naar een nieuw engagement* (Towards a new commitment to society) [1]. It is to serve as a framework for new policies, one of which being that sustainable development should play an important role in research and education:

> in considering the problems of today and tomorrow, stimulating sustainable development is one of the most important contributions our university can make. The principle of 'sustainable development' will have to receive an important place in education and research. Among other things, this means that sustainable development will be an important criteria in formulating a policy on technological development. [2]

As a result of this new commitment, the new curricula of all Faculties are to be evaluated with respect to this criterion. The policy statement on the relevance of sustainable development is an interesting development. The motivation for it is equally interesting. One might expect the find high ideals behind this policy - some form of stewardship of the earth, a challenge to be taken up by science?

The reality is (rather explicitly) that this commitment is a necessity to survive as a university within a society in which public opinion is very much aware of the failures of technology in the past. This type of motivation is more closely related to that of a high-tech company (*improve or perish*) but is also (more positively) part of a trend towards responsiveness instead of only responsibility. In other words, a company should do what is both good for itself as well as for society and should also lead society in this field instead of following trends.[3]

In a scientific community, this may well result in adversarial reactions, since in general scientists are weary of ethics, which are (by nature) not free of value judgements. Engineers, who are not merely content with observing and explaining reality, but who are involved in changing it according to certain goals, cannot do so without ethical choices.

A future engineer who plans to assist in solving society's (technical) problems certainly has to be prepared for discourse on such matters. Education has to provide for this. It is sensible to direct the effort in education at those problems that graduates are most likely to face and/or are

180

most difficult to solve.

A new 5-year curriculum

The cause for the current revision of the course-programme is the re-introduction of a 5-year programme to replace the 4-year programme introduced in 1982. Since its introduction, the average time required by students to complete the programme has changed little and is still over 6 years. This is partly due to the fact that the old 5 year study was compressed into 4 years without (significantly) removing courses or material. This means that the programme was demanding and also that the inevitable delays in the progress of a student are very difficult to compensate for within a given year.

At the same time, there is a general debate in the engineering sciences about problems regarding the relationship between the content and form of the courses, to the real-life problems that the engineers (once they are working in the public and private sector) are required to solve. It is a general view that the universities do not train the students sufficiently in problem solving abilities and design skills. (This is just one of the remarks; others include the question of managerial skills and internationalisation.) At the same time, a discussion on the future of geodesy in the Netherlands has taken place, which although it has not led to consensus [4], has had some impact on the content and structure of the new curriculum.

Goals and restrictions

An increase in the length of educational programme, from 4 to 5 years, was granted by the Minister of Education under the condition that a high proportion of students will actually be able to complete in 5 years. In the revision of the programme, the University has added the issues of sustainable development, a closer relationship between the content of the programme to the professional field, problem solving abilities and design skills. This together has meant a major restructuring of the programme.

The career-paths of geodetic engineers

As a reaction to the lack of foundation in the discussion on the future of geodesy, a survey was held by the Faculty of Geodetic Engineering to trace the career-paths of our graduates [5]. The results helped give a better indication to the Faculty and to the students of the type of geodetic problems the graduates will be faced with.

The survey shows that 81% of those who graduated since 1960, consider their present work to be within geodesy. 58% of this work can be found in land-information (collection, processing and distribution of geo-data), 31% in land management (planning) and 11% in fields usually referred to as higher geodesy (gravity and satellite measurements, etc.). A little over half the jobs are in the public and semi-public sector. A high proportion (47%) have moved into managerial functions (including project management, usually within around 5-10 years after graduation), whilst 27% work in consultancy and research and 9% in education (including non-geodetic teaching such as mathematics). A small number of graduates work (or have worked at some time during their career) in other countries than the Netherlands. Of these, the geodesists have mostly worked in research and education at universities or assisted in the establishment of cadastral registrations. Consequently, there is little reason to over-emphasise the international component of the curriculum (students with interest in international development or implementation of cadastral systems can take specialised, voluntary courses on these subjects), especially since there are no clear indications that this situation will significantly change over the next few years.

Relevance of geodetic subjects to sustainable development

It might be argued that several fields of study in geodesy are not directly relevant to sustainable development. This is true for gravity and satellite measurements, deformation measurements and for land-surveying. However, both research and managerial positions will involve contact with data-users who have specific questions that will have to be answered on the basis of land information. Examples might include land subsidence from oil and gas drilling and sea-level rises as a result of climatic change. Awareness of these environmental problems and the terminology involved in them will be required. The connection between these subjects and sustainable

182

development is, to a large extent, indirect.

For land management and planning the connection is more direct. From the geodetic perspective (as it is defined by our group in Delft) the emphasis is on the informational, legal and financial means of *change in the relationship between human and land* as a result of implementation of land policies. Spatial planning, in a country where 100% of the area is as intensively used as in the Netherlands, involves changing this relationship. This applies not only to the direct relationship of the present owners (or users) with their land, but also to that of neighbouring owners and users. Change also has a wider implication for the rights of all parties concerned; their rights entitle them to the implementation of the plans with appropriate care (and compensation).

This points to two themes in spatial development that can play a role in education:

1.   *A most connected theme: urban and rural development*

Historically in **rural areas**, land development (and in particular rural land reallocation) is a discipline of geodesy. There is room for broader attention to rural subjects, because of the growing complexity of development projects.

The role in **urban areas** is growing because of the trend, in recent years, to develop urban projects in multi-disciplinary teams (for instance under direction of an urban designer). This means that financial and legal consequences are addressed much earlier and can have an influence on the design. The purpose of this approach is to accelerate the development projects (and waiting for hidden problems to appear at the implementation phase).

2.   *A theme connected to the largest employment*

The field of largest employment will bring the most graduates in contact with the subject. Geo-information is extensively used in planning processes. What types of information are needed and what quality of information is needed?

Surveyors have a tendency to go for the highest possible accuracy that the available instruments will allow (although in theory the degree of accuracy is dictated by the application). Planners, on the other hand, often require relatively low accuracy and are more interested in aggregate information.

Education in geo-information (either in this context or otherwise) should, therefore, pay due attention to aggregation-methods, quality-measures and error-propagation of both administrative and geometric information.

The second theme will not be explored further in this paper. The first theme is expanded upon in the next sections.

Sustainable development as an educational tool

Apart from these directly connected themes, there is another reason why sustainable development can play a role in education: problems regarding spatial design and development and especially in connection with the environment are *hard* problems. This is, of course, due to their multi-disciplinary nature and the uncertainty of the (future) results. Therefore, the environment is a good subject for the teaching of integrating, problem-solving abilities and design skills.

Sustainable development in spatial development

In the spatial development of the Netherlands, some difficult choices will have to be made. Demographic and social changes continue to demand more housing, more recreational facilities and more light industrial areas. Technological change might give different patterns of use in the future (telematics) but opinions are divided as to its impact. Increases in mobility are continuing, as Figure 14.1 (taken from a recent government policy plan) indicates, and this threatens the accessibility of cities since traffic congestion is on the increase. Furthermore, even with an increased environmental awareness, decisions with a negative impact on the environment are continually necessary and compensating measures must be sought.

Although these forecasts are not cast in concrete, they indicate some clear trends with which it is possible to work in planning processes. Building good models is still necessary because the limited accuracy is a risk to the credibility of the planning process in the long term. This does not mean that

184

the models should be long term. At some point, the inherent chaotic, non-linear nature of reality [6] makes long term prediction of the behaviour of such systems very difficult or even impossible.

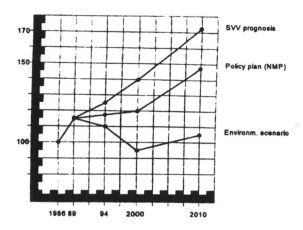

Figure 14.1   Scenarios for the development of kilometers by car in the Netherlands in million kilometers

In the political process, this means that politics should neither overburden society with measures nor be afraid to demand what is necessary to reach short-term goals. It should certainly define what is necessary to change the directions of a negative long-term development.

In Dutch society, which tries to decrease the emphasis on hierarchical policies in favour of networks of groups and responsible citizens, this means that individual choice (also of limited resources) is promoted. This approach, however, should be balanced by confrontation with the consequences of those activities (for instance, in the form of an eco-tax) to promote a change of attitude. In fact, this can be regarded as a translation of the economic principle - that limited resources are expensive.

In the field of spatial development, the government (national, provincial and local) can discourage negative developments, like the growth

185

of mobility by thoughtful design of urbanised areas. Not only by providing environmentally friendly forms of transportation in sufficient supply, but also by making access to roads (a limited resource) more expensive.

Communication

The planning process is very much a communication process. There are some good examples of planning processes that are accompanied by good communication plans. There are however also examples of projects that have failed due to persistent misunderstandings of the goals that were to be achieved.

Bad communication is visible between the agricultural sector and nature preservation organisations [7]. Agriculture is commonly blamed for dehydration problems, pollution and disruption of nature and for overproduction. On the other hand farmers have little confidence in the necessity of nature-improvement measures. This is a general picture, examples to the contrary are also available.

The attitude of the farmers is certainly partly explainable from the state of siege they feel themselves in due to the barrage of environmental measures that they had to deal with over the last years, including a ban on live-stock increase in certain areas, due to ammonia-pollution.

Implications for education

Until recently, the emphasis of the Faculty on education has been very much on the legal and technical side of spatial development. Students should be made aware of the communication side of these processes. This will increase their ability to function in a multi-disciplinary environment and their ability to 'get the job done'. On the technical side, the role of models in education should be changed from purely mathematical models (in for instance surveying) to heterogeneous and dynamic (growth) models [8].

Urban planning

Usually the term 'durable' is used in combination with sustainable development [9]. Solutions generally have a limited life-span, after which the problem needs to be re-solved, under different circumstances. Attention to the durability of solutions promotes sustainability in this particular sense, in that it takes longer before a disinvestment in removing and re-investment in re-building has to take place.

Extra effort is often required to achieve a durable solution and it can be a problem to convince all concerned that it is worthwhile. The extra short-term (financial) effort may be off-set by savings that are made in running-costs, for instance in energy-consumption. In the long-run, less resources are required and this long-term thinking is exactly what needs to be promoted.

For instance, an ecological way of building houses has the reputation of being expensive. Research has shown that it is not [10]. The building industry has not been at the forefront of new ideas, partly because of the lack of need to change.

Attitudes towards an environmental way of building are changing, partly because of conviction on the part of the builders, partly because of the regulations put on activities on building sites (so on the process, not on the product!) and not enough because of governmental encouragement.

Design of lay-out plans

Learning from examples is especially appropriate in fields that are rapidly evolving and where design principles are hard to come by. Some effort is being made of cataloguing good examples of lay-out design. Exposure to such examples stimulates new solutions.

An example, given below, illustrates how the traditional design in the lay-out of an urban area can be transformed:

> houses are usually longer than they are wide, and are built with the shortest side turned to the road in front. One way of decreasing energy-consumption is building houses turned with their longest side exposed to the sun [11]. The introduction of such an idea upsets traditional thinking and requires some flexibility on the part all those involved in the planning process.

Implications for education

In education, the consequences of the need for durable solutions can be made clear through exposure to good examples of sufficient complexity and to the application of techniques such as cost-benefit analyses on such examples. This can also include the use of environmental impact studies in order to learn to interpret and use these studies in the planning process. In general, the ability to handle heterogeneous and qualitative data is an important asset for any problem-solver (and so requires some change in a quantity-oriented engineer).

Rural planning

One of the factors in the growth of agricultural production, has been the rationalisation introduced by land consolidation. This process has also meant the loss of many natural values in rural areas. Now, after broadening its scope, the same procedure is used to realise natural and environmental goals. This has maintained the interest, in what is now called 'land development', at a high level although conversely, the popularity of this policy instrument has decreased among farmers for the same reason.

Land development, with its abilities to free land in the required places for new functions, through the use of exchange of land-use rights, remains the most powerful tool for realising spatial development goals.

Design process

Planners usually prepare policies on the basis of sketches, basic concepts, etc. In the implementation phase of projects derived from such policies, other issues emerge. Ecologists, for example, use the concept of indicator-species to characterize an eco-system. An example of this approach and some consequences for lay-out design are described in an article by *Elzinga et al* [12] from which Figures 14.2 and 14.3 are shown here.

In theory, it is possible to take (spatial) measures for the benefit of an indicator-species, which will also benefit other species in its food-chain. The indicator-species, therefore, also serves as a convenient aggregate entity for the ecological goals in an area. This is both a good planning strategy as well as a clever communication strategy to mobilise public opinion.

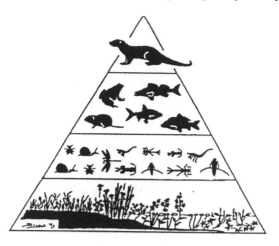

Figure 14.2 The otter is at the top of the food chain. Adoption of the otter as an indicator species will give rise to spatial consequences

In land development projects, the use of this concept gives concrete goals for the realisation of living-areas and -conditions, corridors, etc. Thus, land needs to made available to realise the measures (Figure 14.3) and it is, therefore, one of the responsibilities of the geodetic engineer involved in a land development project.

Implications for education

The design process is heavily influenced, on the one hand by limitations of a legal nature, financial restrictions etc., whilst on the other hand by the many project objectives, which need to be taken into consideration. Satisfying the latter, within the framework of the first, is a major challenge.

Knowledge of the application of legal and procedural elements in changing land use is required in order to implement the policies and the measures. In addition to land law and project planning, this includes knowledge of the concepts that are used by other disciplines in the project, such as ecology.

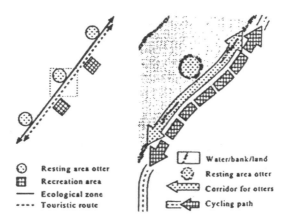

Resting area otter
Recreation area
Ecological zone
Touristic route

Water/bank/land
Resting area otter
Corridor for otters
Cycling path

Figure 14.3 After translation of the concept (left) into a spatial structure (right), land requirements become apparent. (Elzinga, 1995)

Conclusion

This paper has set out to connect the topics of sustainable development and spatial development in a Dutch policy and educational context. The curriculum, which started in 1995 contains some completely new courses, whilst others have been adapted. This is already giving new opportunities to realise some of the points discussed in this paper.

Notes

1. College van Bestuur, 1994. Naar een nieuw engagement, een strategische visie voor de TU Delft. The statement is a starting point for discussion and for formulating new policies.

2. From 'Naar een nieuw engagement, een strategische visie voor de TU Delft', Mission and goals, page 9. Translation by the author. Original text: "Beschouwen wij de problemen van vandaag en morgen, dan is het stimuleren van 'duurzame ontwikkeling' één van de belangrijkste

bijdragen die onze universiteit kan leveren. Het uitgangspunt 'duurzame ontwikkeling' moet een duidelijke plaats krijgen in onderwijs en onderzoek. Dit betekent onder meer dat bij het formuleren van een technologiebeleid voor TU Delft duurzame ontwikkeling een belangrijk criterium is."

3. Kouwenhoven, A., Inleiding in de economische ethiek, 1989. Callenbach, Nijkerk. Kouwenhoven describes a shift in emphasis in ethics in the field of economy from individual entrepreneurial to organisational level, from acting correctly in individual dealings to acting correctly towards society. At the same time governmental and educational institutions regard themselves more like independent entities that have function like business.

4. Although the report on this subject, 'Wat is waar? Nationaal Geodetisch Plan' (What is where? National Geodetic Plan), by a subcommittee of the Netherlands Geodetic Commission (1995) is a tolerable description of the present situation, it gives little or no analysis of the future 'market' for geodesists. Consequently it is of limited value for the shaping of a course-program that intends to train engineers for this market. Many reacted quit strongly and with a more realistic market view, e.g. Grensverleggende geodesie (Moving boundaries with/of geodesy), J.C. Anneveld, R.J.G.A. Kroon, P.A.G. Lohmann, Geodesia, 1994, 11, in which the authors, among other things, advocate that creativity and entrepreneurship should be taught as a basic attitude.

5. Rietdijk, M., Rosman, F.B., de Wolff, H.W., Wie is waar? (Who is where), Geodesia, 95, 5. App. 85% of all graduates since 1960 were surveyed on their employment and subjects of these positions since their graduation. The 1995 survey shows that over 40% of the current job-market has been created in the 10 previous years. It shows further that there is hardly any unemployment but there are some signs of saturation in the last few years in the existing types of jobs.

6. See for instance the paper Chaostheorie und Umweltschutz - der nächste Schritt? by G. Leidig at the 19th International Symposium of the European Faculty of Land Use and Development, München, 1993.

7. Bos, E.G., van Woerkum, C.M.J., Planning is communicatie. Landinrichting, 1995, vol 35, nr 1.

8. "Most people tend to extrapolate. If the birth-rate is declining for years, you are tempted to think that it will continue. A thinker on the future should be aware of the possibility that it may rise. If all of the intellectual elite thinks that the unification of Europe will continue, linear thinking, then a thinker on the future should point out to them that it might go the other way." Schoonenboom in 'The future of yesterday' by W. Oosterbaan

in NRC, September 30th, 1995.

9.   Provincial Authority Utrecht, De toekomst getekend; perspectievenschets Ruimtelijke Ordening - 2015,1991. The report is a mid- and long-term vision on spatial development of the province. In the document general principles are set out, e.g. concentration of urban development in certain nuclei and along existing infrastructure, instead of · allowing suburbanisation, to preserve the provincial identity in the long term.

10.   P.Broos, Ecologisch Bouwen. Delft Integraal, april 1994. Interview by P. Broos with prof.ir. Wiek Röling of the TU Delft.

11.   Ibid.

12.   Elzinga, G., Van Tol, A., Philipsen, J. 1995. Groene netwerken voor natuur en recreatie. Landinrichting, 4.

15 Spatial planning - a key to sustainability?

G. WEBER

Introduction

In 1996, the Austrian Institute of Spatial Planning, in its study on 'Sustainable Regional Planning', commented that

> spatial planning and regional development have remained strangely untouched by the broad disciplinary discussion on sustainability, and it even seems as if the concern for shaping the future does not apply to the two disciplines.

Although this statement is certainly justified, the opposite is equally arguable, i.e. many of the guiding themes of regional planning over the last years for prevention, have only now been taken up by other disciplines under the heading of sustainability and have thus been 're-thought'. So much is certain, that in the interdisciplinary discussion initiated by the issue of sustainable development, spatial planning cannot remain at the fence. It should, however, play a central role as a mediator between the abstract programme of sustainable development, on the one hand, and the concrete instructions for spatial actors on the other. The link between sustainability and regional planning is a goal-means-relationship.

Why this goal-means-relationship should be characterised by very close relationships can *a priori* - without discussing details here - be reasoned along the lines introduced in the following sections.

Spatial dependence

The concept of securing sustainable development is a behavioural codex for human actions in order to guarantee the 'quality of survival', for mankind in the long-term (Diller, 1996). Since man is an earth-bound creature, the question naturally depends, to a very great extent, on how man organises himself in space both as an individual and as a part of the collective. To develop and to also accomplish acceptance rules for a spatial organisation with a promising future on various spatial scales, is predominantly a task of spatial planning. Sustainability and spatial planning thus meet in their intrinsic anthropocentricity and the resulting spatial dependence.

Cross-sectional orientation

The concept of sustainability is based on the insight that environmental problems cannot be solved when detached from economic and social developments. Thus, it requires that both society and the economy have to be restructured towards long-term accordance with the natural environment (Sustain, 1994). To put it in a slightly differently way; the paradigm of sustainability aims at optimising the interaction between nature, society and economy with regard to preserving the functions of nature. Thus, it goes along with - contrary to conventional environmental protection - an integral approach to resolving these problems.

This cross-sectional orientation, as both a problem approach and a solution, has originally and still remains the point of departure for all spatially relevant planning actions. Its range of tasks is described by legislature as follows:

> Spatial planning aims at the best possible use and securing of habitat in the interest of the common good and thus considers natural conditions, assessable economic, social, health and cultural needs of the population as well as respecting the freedom of the individual development needs within the community. (Spatial Planning Act of Salzburg).

Because spatial planning comprehensively links spatial functions and tries to manage them, despite their obvious contradictions, it is in principle predestined to offer its services in putting the target system of sustainability into practice. Spatial planning does not principally operate on a project basis but follows a system-oriented approach, which suits the guiding theme of sustainability.

Small-scale context

Diller, (1996) commented that

> Sustainable development has to be put into practice on all spatial scales and has to be made the central principle for the actions of all actors and institutions. However, particularly suitable scales can be identified.

These highly predestined spatial reference units are the regional and local dimensions. Arguments in favour of the motto, 'think globally, act regionally, respond locally', are *inter alia*.

Due to the highly complex interactions between the economy, society and ecology, the concept of sustainability will only become grasped and directed, in a comprehensible spatial context, in all phases of conversion, i.e. the starting from

> problem definition and - analysis, through target formulation, selection of measures and - implementation to control of success, (Kopfmueller, 1995).

Moreover, it has to be considered that the acceptance of sustainability rules can only be attained by the principle of proximity. This means that actors can most likely gauge the importance of their efforts and contributions in the dimension of easily comprehensible areas, i.e. at the level of the region, the municipality, the city district responsible, the village and the personal environment. This is particularly true because, in a small-scale context, the degree of identification and the resulting preparedness to take action by the community representatives and also the citizens, is usually the highest.

Considering these aspects, spatial planning presents itself as a mediator for sustainability. It has meanwhile accumulated experience with regard to integral actions in easily comprehensible areas over several decades, and is established in the political-administrative system on both the regional and local levels.

Variety of strategies

Through the comprehensive requirements of the concept of sustainability, i.e. the goal to steer all economic and social areas towards this target, it becomes necessary to exhaust all options of authority in attaining this goal.

Even from this perspective, spatial planning has been developing instruments and strategies, which might be contributed towards realising the goal of sustainability. With respect to sovereign jurisdiction, spatial planning covers a highly differentiated system of planning types for various tasks together with spatial reference levels with which rules and prohibitions of spatial uses may be ordered.

In addition to this authoritarian top-down planning hierarchy, further partnership-oriented bottom-up strategies have been developed over time. The latter give impetus to the level of co-operation between the public, non-governmental organisations and the citizens by emphasising dialogue and financial incentives. In particular, the positive experience within endogenous regional developments at the supra-regional scale, contrasting with village renewal and soft urban redevelopment at the local scale, offer suitable strategies for the mediation and accomplishment of sustainable action.

Prevention orientation

The concept of securing sustainable development has been borne out of the recognition that environmental protection, as an act of 'repair' of the environment, can neither stop the continuing waste of natural resources nor ward off, to any acceptable extent, the increasing environmental degradation with pollutants and noise, but also optical environmental pollution. 'End-of-pipe' techniques have not been able to actually release man and nature out of the dilemma of 'too much resource input / too much waste output'. Drawing on this insight, the concept of sustainability puts its stake firmly on the maximum avoidance of space, energy and material consumption. Mono-causal 'after-care' is replaced by the demand for multi-causal 'prevention'.

This guiding theme may just as well be linked with spatial planning because planning generally means the preventive avoidance of foreseeable conflicts. This is why spatial planning is particularly well equipped - under the pretext of fostering sustainability - to *a priori* avoid conflict-inducing patterns of spatial use. Since spatial planning sets up highly persistent spatial structures (buildings, technical infrastructure) its range of action is intrinsically limited as far as structural changes are concerned. Thus, the strength of spatial planning lies principally in prevention rather than in 'after-care'.

Resource intensity

Since the concern of securing sustainable development is the central issue, methods of reducing the future consumption of natural areas, energy and material must be investigated. Some industrialised countries (e.g. Japan, Canada, Germany and Austria) have conducted material flow analyses to be actually able to identify so-called 'resource gobblers'. According to all reports, the major share [the range is from 70 to 80% (cf. OeIR 1996)] of energy and material input, and thus also of space consumption, is consumed by 'settlements' and 'transportation'. Conclusively, the necessity of saving resources as well as the respective savings potential is generally highest in these sectors.

This insight places spatial planning in a prominent position in the debate on sustainability. Spatial planning, to a high degree, determines where and how buildings and technical infrastructure are located, powered, linked to traffic networks, utilised, repaired, and eventually disposed of. It also has a decisive say in defining where, and under which spatial conditions, the unbuilt and principally natural areas may be kept to secure food, replenish renewable energy resources, and in particular the ecological structure. With this range of tasks at hand, spatial planning is at the command of future-oriented resource management and thus carries the responsibility for putting this concept into practice.

Securing quality of life

The concept of sustainability not only aims at a renewed way of handling resources and space, but also to a new approach that can be termed 'the quality of life'. The goal is a dematerialisation of 'the good life', and a decoupling of welfare and affluence.

Also in this context, spatial planning may contribute substantially because it not only concerns the purely quantitative coverage of physical spatial elements to satisfy basic human needs, but also increasingly has a simultaneous affect of reaching qualitative goals in the sense of creating comfortable living space. Undoubtedly, spatial planning is called for to rethink the concept of what is regarded as desirable and to adapt to the requirements of sustainability. Since such revision cannot simply be decreed, new ways of implementing this change needs to be examined. Thus, not only the final results but the existing decision processes need to be critically examined regarding their utility.

Value bonds

The concept of securing sustainable development is an attempt to 'create intergenerative justice through preservation of the resource stock' (Bundesforschungsanstalt für Landeskunde und Raumordnung, 1996). With reference to this target definition, three central requirements have been formulated whose subtargets and actions have to be satisfied. These are the ethical criteria of justice, efficiency and sufficiency.

The basic idea of the concept of sustainability is (besides the intergenerative aspect) also the global compensation of interests between north and south. 'The surplus here and the need there' (Bundesforschungsanstalt für Landeskunde und Raumordnung, 1996) demands more equal patterns of distribution of resource consumption on a global scale.

It is the task of spatial planning now to point out that, despite this global view, the postulation of justice has also to include a small-scale view. Thus, it must be concerned (with the necessary changes in the economy and the society) that rural areas do not continue to lose importance as living and economic spaces relative to agglomerations; that a stable regional urban-rural-network remains intact and developed responsibly, and that quality of life (not in the sense of owning a lot but living comfortably) is guaranteed in all parts of the country to a high level. This is where the commendable, already existing strategies developed by spatial planning, i.e. endogenous regional development, participatory town planning and village renewal, need to be firmly established as important steps towards putting the framework of the holistic concept of sustainability into practice. Through these avenues, it can be agreed upon by means of dialogue in a concrete spatial context. Examples might include discussions of how the regional economic circuits can be strengthened, how more autonomous and more ecologically sensitive regional energy supplies should be developed, how self-sustaining regional networks could be set up and what the spatial consequences of these new concepts should be.

Another cornerstone to the concept of sustainability, is the requirement of efficiency, which refers to the economical treatment of materials, energy and space. In addition to the technological changes serving these goals, the corresponding organisation of space in this context is of crucial importance. Here spatial planning is called upon to accomplish, with a greater degree of resolve than in the past, its 'principle of short paths'. Short paths in the regional and local context, go hand-in-hand with time savings, economical energy, material and space consumption, keeping room for dispositions - even in a more remote future, and with increased quality of life. On a supra-regional scale, the aim is to run against the tide of regional and functional segregation and centralisation trends (the opposite is termed 'sprawled clustering'). On a local level, daily errands and housekeeping should be possible in 'Umweltverbund', i.e. on foot, by bike, or with public transport. This can usually only be achieved though moderate density and functional mix of congruent land uses.

Another central requirement of the concept of sustainability is sufficiency, i.e. abandoning further resource consumption either to avoid foreseeable or assumed ecological risks, to reduce existing risks, or to meet the postulation of justice.

When it comes to the contested question of 'how much is enough?', spatial planning can again offer its services (however, not always successfully so). It, nevertheless, is the nature of this task to set up control mechanisms by ordering spatially relevant laws against conversion of uses.

Openness

The concept of securing sustainable development is not to be regarded as a finished theoretical construction but as an iterative process, heading in the direction of long-term survival. Accordingly, it is principally open to changing preferences and new scientific insights. This belief in openness refers, not only to changes over time but also to differentiations within a spatial context. Regional and local differences are to be respected as the framework of the concept of sustainability. Diversity is a strength and requires situation-specific development strategies.

Furthermore spatial planning has, since the seventies, increasingly evolved towards openness.   Thus, the times of long-term, static-authoritarian, spatial development guidelines in the form of expert plans for the people 'affected', which were quasi-ordered by the agency of this discipline, are gone. Today the quality of spatial planning is, to a lesser extent determined by detailed short-term development targets put forward in a programme or plan, but increasingly by the positive changes that the planning process has on the awareness and actions of the planning participants. Thus, it is important to the degree and in the form in which the population is involved in the planning process, and with how much acceptance the planning product is met with by the local population. This will, hopefully, lead to an active participation by the local community in putting the content of the plan into practice, and if a continued discussion of the further development of the planning content takes place, and so forth.

Complementarity

The link between sustainability and spatial planning should be regarded as a symbiotic relationship, as a relation where the weakness of the one component may be 'intercepted' by the strength of the other, and a relationship where the whole reflects positively back on the individual parts.

Thus, we have to state that spatial planning today has undoubtedly lost some of its initial thrust and fascination with regard to social and political awareness. The initially high expectations have been replaced by a rather sobering balance of day-to-day failures (e.g. the crawling sprawl, which spatial planning always maintained to restrain). In short, the superior objectives of spatial planning of serving the long-term public interest have been sacrificed to, sometimes, short-term egotism in many individual decisions. The latter has obviously the more convincing arguments on its side. This is also the case because spatial planning has lately been lacking in visionary strength.

The weakness of spatial planning today may be resolved by adopting the paradigm of sustainability. It will force planning authorities to, once more, interpret 'planning as a mental discussion with future', (Lendi, 1994) and this may again give spatial planning more power and thrust.

Conversely, the concept of sustainability is now a conglomerate of visionary thoughts that literally searches for a foundation in firm ground. To transform its central requirement such as dematerialisation, deceleration, identity, diversity, naturalness, etc. into concrete patterns of activity, will be a central task for spatial planning. Within this framework, it offers the advantage of being able to easily convert abstract demands into concrete goals and actions within a particular spatial context.

Conclusion

The phrase, 'Suit the action to the words', may be rightly employed at this point. The parallels between sustainability and spatial planning discussed in this paper may not have any effect if thoughts are not followed by firm, spatially relevant actions. Thus, it will undoubtedly need increased efforts by participants in the future, particularly in the field of spatial planning, to actually close the gap between theory and practice, i.e. between words and actions.

References

Baccini, P. 1994. Vom Umweltschutz zur nachhaltigen Ressourcenwirtschaft. In: *Die Zukunft beginnt im Kopf.* Hochschulverlag an der ETH, Zurich, Switzerland.

Bund und Misereor (Eds.) 1996. *Zukunftsfaehiges Deutschland. Ein Beitrag zu einer global nachhaltigen Entwicklung.* Birkhaeuser Verlag, Basel, Boston, Berlin.

Bundesforschungsanstalt fuer Landeskunde und Raumordnung (Ed.) 1996. *Staedtebaulicher Bericht. Nachhaltige Stadtentwicklung. Herausforderungen fuer einen ressourcenschonenden und umweltvertraeglichen Staedtebau.* Eigenverlag, Bonn, Germany.

Fuerst, D. 1986. Oekologisch orientierte Raumplanung - Schlagwort oder Konzept? *Landschaft + Stadt,* 12/1986.

Kopfmueller, J. 1995. Ungeloeste Probleme der Sustainability - Leitidee. In: *Nachhaltigkeit aus naturwissenschaftlicher und sozialwissenschaftlicher Perspektive.* Hirzel Verlag, Stuttgart, Germany.

Lendi, M. 1993. Ethik der Raumplanung. In: *Planung als politisches Mitdenken.* Verlag der Fachvereine, Zuerich, Switzerland.

Lendi, M. 1994. Rechtliche Moeglichkeiten und Grenzen der Umsetzung des

Nachhaltigkeitsprinzips. *DISP,* 117.

Lucas, R. 1995. Die Kreislaeufe schließen: oekologisch Wirtschaften in der Region. In: *Toblacher Gespraeche.* Eigenverlag, Toblach, Italy.

Minsch, J. 1994. Agenda für eine "Nachhaltige Entwicklung Schweiz". In: *Bulletin,* 253.

Moser, F. 1994. *Inseln der Nachhaltigkeit?* Manuskript, Graz, Austria.

Moser, F. 1995. *Grundsaetzliche Ueberlegungen ueber die derzeitigen Moeglichkeiten der Umsetzung des Prinzips der Nachhaltigkeit.* Manuskript, Graz, Austria.

Oberoesterreichische Umweltakademie (Ed.) 1995. *Durch nachhaltige Entwicklung die Zukunft sichern. Landesumweltprogramm fuer Oberoesterreich.* Eigenverlag, Linz, Austria.

Oesterreichisches Institut für Raumplanung OeIR (Ed.) 1996. *Nachhaltige Regionalentwicklung, Teil 1.* Eigenverlag, Wien, Austria.

Reith, W.J. 1985. *Raumordnung ist Umweltvorsorgeplanung.* Manuskript, Wien, Austria.

Spangenberg, J.H. 1995. *Towards Sustainable Europe. Zusammenfassung einer Studie aus dem Wuppertal Institut fuer Klima, Umwelt, Energie.* Eigenverlag, Wuppertal, Germany.

Sustain (Ed.) 1994. *Leitfaden zur Projektbeurteilung nach dem Gesichtspunkt der Nachhaltigkeit.* Manuskript, Graz, Austria.

Weber, G. 1994. Fuenf Thesen zur oekologisch orientierten Raumplanung. *Raumordnung aktuell,* 3/94.

Weichhart, P. 1995. *Die "regio salisburgensis" oder "Statt-Planung in Salzburg".* Manuskript, Salzburg, Austria.

Wuppertal Institut fuer Klima, Umwelt, Energie (Ed.) 1995. *Zukunftsfaehiges Deutschland. Kurzfassung. Ein Beitrag zu einer global nachhaltigen Entwicklung.* Eigenverlag, Wuppertal, Germany.

16 Integrated agrarian structures and regional development

E.C. LÄPPLE

Introduction

In the Federal Republic of Germany, the agrarian structure is undergoing change. This change is not only affecting agriculture and forestry, but also having a far-reaching impact on the use of land and is changing the functions of rural areas and their villages.

Two events, in particular, have changed the sustainable framework conditions for the agrarian structure; the unification of the two German States with their different types of societies, and the reform of the European Common Agriculture Policy (CAP). The adjustment of the socialist land law and the collective farming in the former German Democratic Republic to the West German social market economy, together with the European Union's (EU) rules of competition, have forced new developments to take place. In addition, non-agricultural demands on land use are gaining in intensity. These are the competing demands of urban development, industry and trade, transport, water, soil protection, nature conservation and landscape management, as well as leisure and recreational facilities.

The EU through the three structural funds of EAGGF, ERDF and ESF is already implementing an integrated policy for rural areas. In principle, the EU funds serve to co-finance national promotional measures, which have to be taken to adjust to EU conditions. Through the promotion of the, so-called, AIM 1 and AIM 5b areas alone, investments of approximately DM30 billion will be made by the year 2001. In addition, further investments totalling DM600 million will be made in the rural areas during this period the EU LEADER and INTERREG initiatives. It is, therefore, very important that urgent consideration be given to identify how European, governmental, regional and communal instruments may be used to improve the agrarian structures within the context of integrated regional development.

Definition of terms

Regional planning

The task of regional planning is to create the most satisfactory conditions possible for people, by providing them with an optimal link to their sphere of living.

Regional development

This is part of the realisation of regional planning. It encompasses the planning, preparation and execution of all measures suited to preserve and improve the living, economic and recreational functions of rural areas in particular. It also provides for the sustained improvement of the living conditions outside the urban areas.

Agrarian structure

Both regional planning and development have an agrarian structure component. The agrarian structure is understood to be the relationship between, on the one hand the production factors employed in agriculture (i.e. land, capital and labour), and on the other hand, production in the agricultural sector. Thus, the definition of an agrarian structure encompasses elements of the natural and cultural environment in rural areas and villages, to the extent that agriculture and forestry determine them. The spectrum of tasks far extends beyond the production of foodstuffs. It includes; securing the natural and man-made landscape, tourism in local areas, and contributes to the preservation of the rural infrastructure.

The economic strength of rural areas

In order to preserve the ability of rural areas to function and to be viable entities, the economic conditions of the framework must be changed. This is especially necessary in those locations where agriculture can no longer be the decisive economic factor. Besides the production of foodstuffs and raw materials, additional sources of income - both inside and outside of the agricultural enterprises - must be opened up, such as in the trade and service sectors. Services in environmental and nature conservation, together with landscape management can be provided as a corresponding component of farming, but cannot be financed through agricultural prices. Those services must, therefore, be adequately remunerated from other sources. Many of those services will rely both upon labour as well as creativity, both of which are available in sufficient quantities in rural areas. That range of services, however, will not be required if the environment is not attractive and there is, therefore, no functioning market. Therefore, one of the key factors in the improvement of the agrarian structure

A key factor in improving the agrarian structure is the 'village', which in Germany is being promoted as an alternative settlement area to living and working in conurbations. Providing the population density is kept at a sufficient level, the basic infrastructural facilities - together with a lively village culture and a healthy environment, can be maintained. All village inhabitants (whether the farming communities or the newcomers) can benefit from this since sufficient inhabitants will ensure that an adequate number of schools are maintained, decent educational and cultural facilities will be provided, shops and commercial facilities (including banks and post offices) will not be closed down. This will lead to an environment that will fulfil the needs for everyone within the area. The catalyst for this, village structure, is the agriculture and forestry enterprises, whose existence will be more indispensable than ever.

Integrated development planning

In Germany, the so-called agrarian structure pre-planning has proved to be a successful instrument in establishing an efficient agricultural sector having a varied structure. It has subsequently evolved from the pre-planning stage, to the adjustment of agriculture and forestry to suit modern conditions. This has proved to an effective basis for the co-ordination of projects aimed at preserving the functionality of rural areas and their villages. It may also be used to indicate possible areas of conflict, to provide opportunities for development and satisfies the need for decision-making within the agrarian structure. A further role for the agrarian structure pre-planning has been in the development of regional land-use concepts and area-specific role models.

The principal advantage of the pre-planning exercise lies in recommending (and under some circumstances, alternative) plans of action for packages of measures, which are not only desirable, but can also be realised. It is, therefore, little wonder that this initiative is coming, in particular, from the new *Laender* in Germany, to develop and expand agrarian structure pre-planning into an agrarian structure development plan. This will be advantageous for the following reasons:

1.   It should prevent the planning process from being determined from the point-of-view of an exclusively agrarian structure, thus neglecting to take into account the potential for the development of the remainder of the region.

2.   The contributions of agriculture and forestry towards the sustainable improvement on the basis of their continued existence are not only to be determined on a sector-by-sector basis, but also as part of an integrated regional development approach.

3.   The possibilities and limits of shaping the future of the regions determined by agriculture and forestry are to be ascertained and set.

4.   It can be shown what legal conditions still have to be created and which financial, administrative and personnel-related resources have still to be brought together for this purpose.

Thus, agrarian structure development pre-planning is to prevent planning, which is related to regional planning, from being implemented without taking into consideration the requirements of the agrarian structure due to a lack of individual, practical knowledge of agriculture and forestry. In many regional planning processes, agriculture and forestry is only recognised as:

1.   a *highly subsidised* producer of *surplus foodstuffs, which are sometimes a risk to human health* as well as a producer of *overly expensive* raw materials;

2.   a *significant* polluter of the environment;

3.   an *inexhaustible* supplier of building land, as well as land for;

   i.   infrastructural facilities (*including the areas for replacement and compensatory measures according to the Nature Conservation and Forestry Acts*);

   ii.   nature conservation purposes (*at least 10% of the total*

206

area of the countryside);

iii.   sporting facilities (*such as golf*).

Conclusion

The image of rural areas has been greatly influenced by the work of farmers and foresters.   Private initiatives by the local inhabitants, supported by communal development measures and government promotional programmes, have raised the standing of rural areas in society. Rural areas are now considered an attractive alternative to life in densely populated conurbations. It would, however, be asking too much to expect that agricultural policies become the sole driving force behind the development of rural areas.

Regional economic policies must be developed, which combine both social and cultural policies of the region and the communities, as well as the initiatives of the population.   Both policies must, however, be adjusted to one another and brought into harmony with a reliable agrarian structure, in the form of an Integrated Regional Development Plan, examples of which may be found in Saarbrücken and Oderbruch.

17 The ecological scores model of Lower Austria - a programme to foster sustainable cultural landscape development?

W. SEHER

Introduction

The present crisis in the agricultural sector - a consequence of a surplus of agricultural production - expressed in dwindling agricultural income and a drastic fall in the number of farms, places an existing by-product of the agrarian economy, an agriculturally shaped cultural landscape, to the centre of attention. In the light of current agricultural practice, how it will operate and cope in the face of increased competition and how will it be affected by the need to intensify means of production? The likely result is that agricultural and natural habitats will by no means be regarded as simple by-products or even left-overs. Conversely, the further retreat of agriculture from land in areas not suited to intensive cultivation, endangers elementary functions which lie beyond the actual sphere of production. Landscape protection functions, with regard to the preservation of recreational areas, or the protection of habitats from natural disasters, may serve as examples.

The problematic nature and far reaching consequences of these developments have led to modifications in agricultural policy objectives away from its primary focus on food and resource production, towards a multiplicity of non-productive agricultural functions. The basic goal of this type of future-oriented agrarian policy is not only to secure cultivation and the stabilisation of farm income, but also for the rehabilitation and improvement of the natural environment in intensively used agricultural areas (Gatterbauer *et al.* 1993).

Following the recently defined agro-political objectives, sector-specific aid programmes have been, and are being, established in accordance with a singular funding objective. These include:

1. set-aside programmes as market-depressing measures;
2. landscape management programmes for the maintenance of extensive cultivation forms in ecologically fragile sites;
3. habitat conservation and creation programmes for the protection and setting up of single habitats. (Paar *et al.* 1993).

Against this background, the Department of Land Consolidation in Lower Austria has developed the Ecological Scores Model Agriculture (abbr. Ecological Scores Model) as an integral base aid model. This is directed primarily towards production-independent direct payments, the ecologisation and extensification of soil cultivation, and to honour existing ecological outputs. The expression, 'integral', may be interpreted in this context as considering the farm's total area of agricultural cropland in the aid programme. The initial consideration was that if there were to be ecological outputs within agriculture, how should they be evaluated and also how, and with which key, might they be compared (Mayrhofer & Schawerda 1991).

Sustainability, with regard to land use refers to the long-term preservation of biological and economic yielding capacity of cultivation sites (Werner 1995), as well as preserving an intact landscape balance, i.e. maintaining the functions of soil, water and natural landscape elements (Schawerda 1990).

Description of the Ecological Scores Model

The Ecological Scores for Agriculture Model aims at quantifying the ecological output of agriculture within a predefined set of scores. The basic principle of the assessment is the relationship of a farmer's ecological outputs, as expressed in the cultivation of grassland and arable land, to the share of land left to landscape elements. Financial aid and compensation payments will be based on these ecological scores.

Basic structure of the assessment plan

The measure of a farmer's ecological output is the extent to which the landscape balance is intact. This may be defined in terms of the form of cultivation in agricultural areas compared to the provision for natural areas

209

within the farmer's land (landscape elements). A comparison of both parameters, based upon a multiplication process, reflects their interaction in the agricultural ecology. The decisive factors of the landscape balance include such factors as; energy and material flows, ecological relationships and networks, and the climate and water balance. These are taken into account not, in a direct and sectoral manner, but in an indirect and systemic manner. The ecological output of cropland areas, i.e. arable land (incl. special forms of cultivation, such as vineyards or fruit plantations), grassland (meadows and pastures) as well as fallow land, are then assessed.

Parameters for the assessment of cultivation form

Differences in the cultivation of arable land and grassland give rise different assessment parameters. Their scores should nevertheless be comparable. In both cases, equivalent parameters are established - both for cultivation intensity and type.

	Parameters for	
	Arable land	**Grassland**
Cultivation intensity	crop rotation	cutting frequency grazing intensity
	fertiliser input	fertiliser input
	crop protection	crop protection
cultivation type	ground cover	
	fertiliser - type and application method	fertiliser - type and application method
	field size	age of grassland

Figure 17.1 Parameters for the assessment of cultivation form (Mayrhofer & Schawerda 1991)

Assessment guidelines for specific parameters

The scoring method for parameters takes into account ecological tolerance levels. Ecologically tolerable, yet landscape management-insensitive behaviour scores zero. Landscape-sensitive effects, i.e. those having positive effects on the demands of the landscape balance as well as agricultural ecology, score positive points, landscape-damaging effects score negative points.

	Score/ha
crop rotation	0 bis +7
ground cover	0 bis +9
fertiliser input	-9 bis +6
fertiliser: type and application method	-6 bis +8
field size	0 bis +5
crop protection	x 0,3 bis x1,0

Figure 17.2 Score board for cultivation form parameters: the example of arable land (Mayrhofer 1994)

The scores of the parameters are summed (the same system applying to meadows and pastures). The sum is then multiplied by the factor of the crop protection parameter.

Assessment criteria for landscape elements

In the Ecological Scores Model, landscape elements are defined as natural (or only marginally cultivated) areas within open fields which are neither arable nor grassland. Landscape elements directly bordering land cultivated by the farmer are considered (Mayrhofer & Schawerda 1991). The assessment criteria are the size of the area together with the 'quality' of the landscape elements. The relationship between the area of the landscape elements and the size of the agricultural area (expressed as a percentage) serves as the basis for calculating the multiplication factor and thus, the actual score. The qualitative standard is taken into account through deductions or surcharges of the actual size of the landscape elements.

Assessment parameters	Assessment criteria
crop rotation	diversity of crop rotation share of renovating crops (sequence: forage plant/oil seeds, grain, row crops) over the past six years
ground cover	type and duration of additional ground cover
fertiliser input	applied quantity of nitrogen applied per cultivation year depending on soil quality and nutrient demand of the crop
fertiliser type and application method	fertiliser type according to degree of water solubility quantity of nitrogen per single application
field size	area size of cultivation unit
crop protection	number of biocide applications per cultivation year

Figure 17.3 Scoring criteria: the example of arable land (Mayrhofer 1994)

Calculating the ecological score

The score of the respective parameters relating to the form of cultivation are resolved (added and multiplied, respectively) and the relationship with the landscape elements is established by the multiplication factor (cf. also the diagram overleaf).

The Ecological Scores Model has been included in the **Austrian Programme for the Fostering of Environmentally Sound and Habitat-Preserving Agriculture (OePUL)** (in accordance with EU regulation 2078/92) during 1995. The OePUL includes a variety of programmes designed to foster environmentally sound land cultivation, as well as providing compensation as an encouragement to farmers for landscape management.

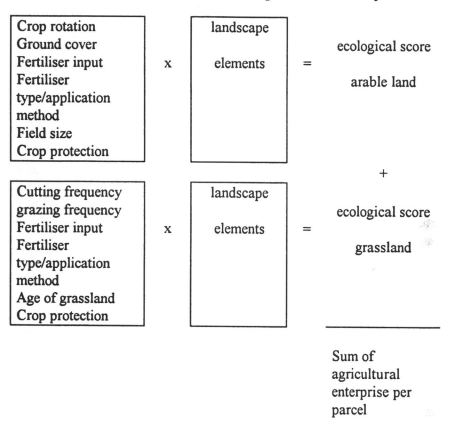

Figure 17.4 Scoring diagram (Mayrhofer & Schawerda 1991)

The Ecological Scores Model: a sound program to foster sustainable cultural landscape development?

The relative merits of the Ecological Scores Model as a suitable instrument for financial aid to encourage sustainable cultural landscape development, will be examined in the light of following criteria:

1. solving agricultural environmental problems;
2. nature conservation;
3. consideration of regional disparities in land cultivation.

Solving agricultural environmental problems

The definition of sustainability of land cultivation refers, to a great extent, to the protection of a stable landscape balance, i.e. the long-term preservation of the functions of soil, water and the natural landscape elements.

Agriculturally induced soil erosion together with nutrient displacement from agricultural land into ground and surface waters, indicate that intensive land use, ignoring site-specific aspects, overburdens the soil and places an increasing ecological stress on water bodies. As a basic model for financial aid for improving the ecological balance of an area, the Ecological Scores Model is regarded as an appropriate concept. It may be used both to assess the situation regarding areas of potential difficulty and to support the necessary steps for improvement.

As a protective action against soil erosion, agro-biological measures such as humus accumulation, ground cover and reduced soil cultivation are equally viable technical measures. These act as de-central engineering barriers to reduce surface water runoff (Schawerda 1993). The proposed classification of protective measures is analogous to the structure of the Ecological Scores Model. The group of agro-biological measures corresponds with the assessment of the form of cultivation, the group of technical engineering-biological measures with that of the assessment of landscape elements. The qualitative and quantitative assessment of erosion-reducing cultivation is possible on the basis of the structure of the Ecological Scores Model. The correlation between the priorities of erosion protection and the assessment criteria of the model has, however, yet to be shown.

The assessment criteria relating to the parameters of crop rotation and the decreasing score from forage plants through to grain and row crops, reflects the risks of erosion from individual groups of cultivated plants regarding ground cover, humus accumulation and cultivation. The importance attributed to the measure of additional ground cover in the Ecological Scores Model takes into account the high erosion risk of uncovered soil outside the period of vegetation cover. The importance of field size for erosion behaviour focuses on the length of erosive slopes. Small sizes reduce the likelihood of erosive surface water runoffs over the long-term and are thus attributed higher scores in the model. Landscape elements such as hedges and borders at right angles to the running direction are central elements of all erosion protection plans in land management and can be enhanced by applying the Ecological Scores Model in conjunction with forms of soil-protecting cultivation.

In the case of nutrient displacement, particularly the leaching of nitrates into the ground water from agriculturally areas, three critical factors are decisive with respect to agrarian land use (Rohmann & Sontheimer 1985):

1. type and duration of plant cover;
2. type and intensity of soil cultivation;
3. type and intensity of fertiliser application.

The factors of plant cover and soil cultivation are indirectly taken into account in the parameters of crop rotation and ground cover. The factors relating to fertiliser application is referred to in the parameters relating to the input and type of the fertiliser together with the method of application.

The influence of individual cultivated plants on nitrate leaching is the result of the type and the duration of plant cover (Kohlmeyer 1985). The significance of soil cultivation lies in the airing of the soil and the hereby fostered nitrification and mineralisation as well as the increased quantity of percolating water (Herrmann & Plakolm 1991). Crop rotation sets the frame for the duration of soil cover and black fallow as well as the form of soil cultivation. The graded score within the cultivated plant group within the parameter crop rotation, from forage plants/oil seeds through grain to row crops, is reflected in the common cultivation practice regarding crop rotation and runoff risk of these cultivated plants regarding type and duration of soil cover and intensity of soil cultivation. Types of additional soil cover, such as planting hibernating intermediate crops score high points and contribute substantially to lowering the leaching potential. Type and level of the actual nitrate fertilisation may be precisely accounted for by the parameters fertiliser input and fertiliser type and application method, and may be scored then. The assessment plan of fertiliser input specifically refers to the nutrient demand of the individual cultivated plants.

In addition to the identification of small-scale groundwater protection areas, blanket coverage extensification of soil cultivation in the catchment area of ground water extraction is frequently demanded (Arbeitsgemeinschaft Flurbereinigung 1993). The resulting cultivation conditions can only become socially acceptable for agriculture in conjunction with targeted supportive actions. Although the Ecological

Scores Model collects and assesses the essential decisive factors for nitrate leaching it can, however, only meet the requirements with supporting information campaigns and awareness building actions for farmers towards forms of ground water conserving cultivation.

Goal attainment with regard to nature conservation

Traditional, area-protective nature conservation has not reached its overall target. The decline in plant and animal species relentlessly continues. One of the major causes lies in the fact that the remaining habitats are small in size and isolated from each other by areas of intensive use (Hasler 1995). A new visionary goal proposes the co-operation of nature conservation and agriculture which - besides large-scale protected areas - includes blanket coverage cultivation extensification and aims at protecting the issues of nature conservation increasingly by the active users themselves. This calls not only for further identification of ecological preference areas but also for an integral financial aid concept which is - unlike many landscape management and habitat conservation programmes - constrained to individual, scattered areas but which takes into account the collective environmental output of an agricultural enterprise (Wurzian 1994).

These criteria, developed by nature conservationists, can be conceptionally fulfilled by the Ecological Scores Model. The linking of cultivation forms with the assessment of landscape elements by multiplication, as well as the capturing of environmental outputs of all agricultural areas of a farm, guarantees the required integral approach. The practical transfer of blanket coverage of nature conservation concepts, however, calls not only for a targeted support instrument but also for accompanying guidance instruments. The necessity of adapting the actions in question to differing agricultural production areas and the various landscape types, highlights the importance of providing guidelines and supportive advice for farmers. Attention should be drawn, in particular, to the steering potential of landscape planning and land management for the sustainable development of rural areas, oriented at ecological principles. Landscape framework plans provide suitable information potential for working out specific conditions for ecologicalisation and extensification measures or nature conservation programmes in the respective regions. The clear objective is the formulation of a local ecological guiding theme, such as linking biotopes or preserving certain landscape elements (Knoll 1993).

Consideration of regional disparities of land cultivation

The necessity of adapting extensification measures to different regional cultivation conditions has already been discussed above. Attention should be given, however, to the regional peculiarities which need to be taken into account in the Ecological Scores Mode. The disparities between areas not suitable to intensive agricultural production and areas suitable to intensive agricultural production are of particular interest in this context. Lower Austria is an excellent example because of its variety of landscapes, both along its north-south and east-west axes. With more general landscape management programmes, the regional peculiarities will be considered to a lesser extent - being expressed in differing terrain and site conditions and the resulting differences in type and intensity of land cultivation. The Ecological Scores Model, whose content and goal are based on data derived in the field, works with the smallest possible unit of the regional financial aid programmes - the cultivation unit. The regionalism thus originates from the farmer himself. The operative conditions of his farm are the yardstick for his personal financial aid programme (Schawerda 1994).

This factor indicates why the Ecological Scores Model will no be capable of levelling the disparities of cultivation intensity and biodiversity between areas suitable and not suitable to intensive agricultural production. The ecological output provided by agriculture, which is by definition higher in areas not suitable to intensive agricultural production, is being honoured accordingly. A farmer upon deciding to participate in the Ecological Scores Model enters the Model with a high score, and his landscape management output by extensive cultivation is honoured fairly. A model taking into account the total environmental output, is distinctly less attractive in areas suitable to intensive agricultural production. Due to low initial scores, the Ecological Scores Model would only become financially attractive when applied in conjunction with additional regional compensation payments (Schawerda 1994).

Initial practical experiences with the Model confirm these ideas. In intensively used agricultural production areas, farmers choose those aid programs which assess only single cultivation parameters and are thus not compatible with the Ecological Scores Model. The Model is in high demand, however, in those regions of Lower Austria with a high share of grassland, small-scale agricultural structures and high provisions of natural remaining areas. With regard to regional disparities, the Ecological Scores Model displays structure-conserving effects which is contrary to the goal of holistic sustainable cultural landscape development.

Summary and outlook

The theoretical concept of the Ecological Scores Model - the topic and aim of this paper - promises to be a suitable instrument for supporting sustainable cultural landscape development. It integrates the central aspects of agricultural land cultivation and combines them into a logical and integral model by joining them by multiplication. Land cultivation, which is targeted at the requirements of soil and ground water protection, may be supported in a target-oriented manner through the guidelines of land cultivation and under consideration of the respective site conditions.

Efforts of putting nature conservation into practice and also going beyond local territorial protection, seem feasible if at the same time models of cultural landscape development are provided. Cultural landscapes are not static states but active processes. Future financial aid programmes can only become efficient if they integrate landscape dynamics within their statement (Suske 1994). The absence of predefined entry barriers into the Model, as well as its lack of distinct phases in its conception, guarantee the dynamic dimension and leaves space for development in land cultivation.

The polarity between areas suitable and areas not suitable to intensive agricultural production will, however, not be decreased by the Ecological Scores Model without additional compensation programmes. Practical experience already shows a development towards a model of rewards for existing ecological outputs by agriculture in areas not suited to intensive agricultural production. The desired incentives for more sustainable cultivation forms in intensive agricultural production areas will, however, not originate from the Ecological Scores Model due to its low financial attractiveness. In the latter, single target programmes are preferred whose cultivation targets are easier to fulfil and which promote higher income from landscape management. These programmes, however, do not meet the integral demand for sustainable cultural landscape development. This emphasises the need for steering instruments, such as actions of land management and the instrument of landscape planning in intensive agricultural production areas, to co-ordinate various extensification and landscape management programmes and unite them towards the goal of sustainable cultural landscape development.

References

Arbeitsgemeinschaft Flurbereinigung 1993. *Landentwicklung-Schutz der Lebensgrundlage Wasser.* Dokumentation und Empfehlungen der Arbeitsgemeinschaft Flurbereinigung, Muenster-Hiltrup, Germany.

Gatterbauer, H., Holzer, G. & Welan, M. 1993. *Agrarpolitik und Agrarrecht in Oesterreich - ein Ueberblick.* Institut fuer Wirtschaft, Politik und Recht der Universitaet fuer Bodenkultur, Wien, Austria.

Hasler, A. 1995. Biotopverbund: Allen Lebewesen Lebensraum geben. *Naturschutz in der Gemeinde,* 3/95.

Herrmann, G. & Plakolm, G. 1991. *Oekologischer Landbau. Grundwissen fuer die Praxis.* Oesterreichischer Agrarverlag, Wien, Austria.

Knoll, T. 1993. Beitraege der Landschaftsoekologie zur Gemeindeplanung im laendlichen Raum. *ZOLLtexte,* 1/93.

Kohlmeyer, M. 1985. *Pflanzenbauliche Maßnahmen zur Minderung der Nitratverlagerung.* Weinheim, Germany.

Mayrhofer, P. & Schawerda, P. 1991. *Die Bauern, die Natur und das Geld. Modell Oekopunkte Landwirtschaft.* Verein zur Foerderung der Landentwicklung und intakter Lebensraeume, Baden, Austria.

Mayrhofer, P. 1994. *Regionalprogramm Oekopunkte Niederoesterreich.* Amt der NOe Landesregierung, Wien, Austria.

Paar, M., Fischer, I. & Tiefenbach, M. 1993. *Landschaftspflegeprogramme in Oesterreich.* Bundesministerium fuer Umwelt, Jugend und Familie, Wien, Austria.

Rohmann, U. & Sontheimer, P. 1985. *Nitrat im Grundwasser. Ursachen-Bedeutung-Loesungswege.* DVGW-Forschungsstelle am Engler-Bunte-Institut der Universitaet Karlsruhe, Karlsruhe, Germany.

Schawerda, P. 1990. Landschaftspflege von oben. In: *Landwirtschaft und Umwelt: Situationsanalyse und Loesungsansaetze.* Ed: Hofreiter, M. F. Schriftenreihe Club Niederoesterreich, Wien, Austria.

Schawerda, P. 1993. *Entwicklungsplanung in den Fluren.* Studienbehelf zur Vorlesung und zu den Uebungen aus Agrarische Operationen, Wien, Austria.

Schawerda, P. 1994. *Die Bauern, die Natur und das Geld. Unterschiedliche Standpunkte, Modellvergleiche und eine Bilanz des Oekopunkteprogrammes.* Verein zur Foerderung der Landentwicklung und intakter Lebensraeume, Baden, Austria.

Suske, W. 1994. *Kulturlandschaft in Niederoesterreich.* NOe Landschaftsfonds, Wien, Austria.

Werner, A. 1995. Entwicklung und Realisierung nachhaltiger Landnutzungssysteme. *Zeitschrift fuer Kulturtechnik und Landentwicklung,* 4/95.

Wurzian, E. 1994. Landwirtschaft und Naturschutz - Partner oder Gegenspieler. *Raumordnung aktuell,* 3/94.

18 The alpine waterscape - sustainable tourism and industrial development in Lago d'Iseo

W.G.O. ZWIRNER

The concept of sustainability has traditionally been applied to either global or regional contexts. Only recently has it also become associated with more specific localities and, in particular, within existing built-up areas. This paper argues for a further local application; the redevelopment pressures in an area of extraordinary scenic beauty and, therefore, strong tourism demand. Lake Iseo in Italy is chosen as a prototype example to argue the case for sensitively balanced sustainable development that takes account of pressures for re-development but also preserves and enhances the physical attractiveness of the locality.

Sustainable development is a concept that draws on two frequently opposed intellectual traditions: one concerned with the limits which nature presents to human beings, the other with the potential for human material development locked up in nature. Sustainable development means more than seeking a compromise between the natural environment and the pursuit of economic growth. It means a definition of development that recognises that the limits of sustainability have structurally and also natural origins (Redclift 1987). An urban or regional scale for analysing sustainability is certainly warranted (Nijkamp & Perrels 1994).

Sustainable development in existing cities is probably best achieved through compactness, i.e. a process of intensification either of built form, such as increases in the density of development, or of activity, such as increases in the use of existing buildings and sites (Elkin *et al* 1991). However, the surveying profession, by its nature, is involved in the destruction of the natural environment and its replacement with a human (built) environment. Its methodologies are the products of neo-classical economics. Sustainability becomes efficient pillaging, but is still pillaging (Eccles 1994, 29).

Lake Iseo (Lago d'Iseo) is the fifth largest of the alpine northern Italian lakes, measuring 24 km long and 5 km wide at its broadest point. At that same point the lake is almost filled by Monte Isola, the largest island of any European lake. Iseo is a tourist lake, its villages are well known for their sailing facilities and their beaches (AA Publishing 1993). Lake Iseo is located between Lake Garda and the east and Lake Como to the west, in Lombardy (Lombardia), which is generally acknowledged to be one of Italy's economically strongest regions. The lake is naturally fed by fresh water from the river Oglio, along the Camonica valley, straight from the alpine summits of neighbouring Engadin in Switzerland. Like other lakes in northern Italy, the water is fresh, rich in fish and very cold!

Administratively, the lake is divided right down the middle, from north-east to south-west, with the western part belonging to the Provincia di Bergamo and the eastern part to the Provincia di Brescia (Lombardia undated). Access by road is well provided on the Provincia di Bergamo side - via Lago di Endine into Lovere at the northern end of the lake - due to the steep rock formation literally vertically dropping into the lake. It is more generously provided for on the eastern side, in Provincia di Brescia, by a dual carriageway eventually of motorway quality - mostly tunnelled. There is a fairly infrequent railway line from Brescia to Edolo but the bus services are reliable and frequent, serving both sides of the lake (albeit run by different companies each of which insisting on their own ticket issue). Passengers can be in Milano or Brescia within two hours every hour or so. Of course, lakes invite public boat traffic and Lake Iseo has a small flotilla of very up-to-date ferry services between Pisogne, Lovere, Riva di Solto, Sensole, Peschiera, Sulzano, Iseo and back. Alas, the time table arrangements are not necessarily suiting everyone. One may go from Lovere to Pisogne in the morning, but will have to wait most of the day to catch the return boat. From Pisogne and Lovere to the stunningly beautiful island of Monte Isola, this is a visit, which is surely top of any tourist agenda on Lake Iseo. The boat timetable offers one morning trip out and an early afternoon trip back.

The main towns and villages on Lake Iseo include Iseo (ca. 9,000 inhabitants), Lovere (ca 6,000 inhabitants) and Sarnico (ca 6,000 inhabitants) and Sulzano (ca. 1,500 inhabitants) (Guide Michelin 1993). They share two principal economic bases: industry and tourism. Primary industry includes agricultural small holdings, including the vineyards of the Franciacorta to the south of the lake, fishing on Monte Isola (AA Publishing 1993, 85), and marble quarries often directly facing the lake shore. Secondary industry includes furniture manufacture and the making of pre-fabricated concrete elements for the construction industry especially along the Camonica valley. In Lovere, there is also a large steel mill. This is due to be closed mainly because its accessibility no longer depends on lake traffic and heavy goods vehicles find it difficult to negotiate the hair-pin curves of the roads. With its spectacular views of the lake and the mountains all the way up to the snow-capped Engadin, this steel mill site must surely rank as one of the most attractive re-development sites in Europe. What development proposals suggest themselves?

The obvious choice would be tourist facilities, including a large, modern hotel, tennis courts and other sports facilities (besides the already existing indoor and outdoor swimming pool nearby). The lakeside location also suggests marina development with mooring for private yachts adjacent to holiday homes. However, local business people in the area are definitely opposing any major development for tourism and especially the construction of a major international hotel (Lab 1995). They argue that the number of bed-spaces are just right for Lovere and beyond and are convinced that the local council (la commune) would fight any major tourism development. Similarly, locals would argue that there is no need to develop new industrial plant on this site. This being Lombardia, there is no unemployment now and unlikely to be in the future.

Under these auspicious circumstances, what would be a sustainable form of development for this peninsula acceptable to both the commune and local people? The solution is probably a mixed-use scheme that combines housing for local people with holiday homes attached to a marina, perhaps one small hotel, a supermarket and a few smaller shops and workshops for craftsmen. There is a need to make the green revolution locally equitable, sustainable and environmentally friendly (Anink & Boonstra 1995) by starting to identify the locally existing socio-economic demands and seeking local approval for sustainable change (Zwirner 1995). Primarily, development urgently needs to blend into the physical fabric, echoing the traditional, and aesthetically pleasing, townscape of Lovere and the visual immediacy of lake Iseo (Cullen 1961). It should be the concern of every potential developer to increase the aesthetic quality of the lake for all users, both local inhabitants and enchanted visitors.

References

AA Publishing 1993. *Italian Lakes*, Basingstoke, Hampshire: The Automobile Association.
Anink, D. & Boonstra, C. 1995. *The Handbook of Sustainable Building: ecological*

choice of material in construction and renovation, London: James & James.

Cullen, G. 1961. *The Concise Townscape,* (1st ed. 1971) London: Architectural Press.

Eccles, T. S. 1994. *The Great Green Con. Surveying and Environmentalism,* Occasional Paper No. 6, School of Surveying, Kingston University.

Elkin, T., McLaren, D. & Hillman, M. 1991. *Reviving the City:towards sustainable urban development,* London: Friends of the Earth.

Guide Michelin 1993. *Italia.*

Lab, A. 1995. Personal communication in Hotel San Antonio, 21st July.

Lombardia (undated) *Carta turistica dei laghi,* Novara: Istituto Geografico de Agostini.

Nijkamp, P. & Perrels, A. 1994. *Sustainable Cities in Europe. A comparative analysis of urban energy - environmental policies,* London: Earthscan.

Redclift, M. 1987. *Sustainable Development. Exploring the contradictions,* London, Routledge .

Zwirner, W.G.O. 1995. Green versus Rural Development - identifying the issues. Paper to the *22nd International Symposium of the European Faculty of Land Use and Development, 8th-10th June, on 'Ecodevelopment of Rural Areas'.*

19 Marinas and sustainable development: making the most out of marinas

D. WILLIAMS

Introduction

Marinas, 'are big business' (Corrough 1989).   Throughout the world ownership of boats is increasing, leading to a demand for more marinas. In the USA, one in fifteen persons own a type of boat (Currough *op cit*). Interest in boating has grown so much that, the space for moorings has become difficult to obtain and this has lead to a greater use of stack storage systems.

The definition of a marina no longer allies itself solely to an area where a yacht can be moored.   The Marina Village is already a going concern, which according to Drumgoole *et al* (1989), incorporates 'increased mooring facilities for pleasure craft and extra land based facilities'.   The adjacent facilities on the land can include hotels, shops, golf courses, parks and water based sports, such as jet skiing and scuba diving.   These developments exist world wide.

An increase in marina developments will not be without their effects on the natural environment.   As Jackson (1984) argues, tourism facilities such as marinas, often 'straddle dynamic and highly vulnerable littoral zones'. Careful attention must, therefore, be given to the means of preserving and protecting the area in which the marina development takes place.   From the outset, it must be clear that the right site has been selected.   Certain sites may be more environmentally sensitive than others when faced with marina construction.   Marinas must be sustainable.   Safeguards must be adopted aimed at protecting the natural environment whilst ensuring the value and the success of the marina is maintained in future years.   Their development must also be seen to provide benefits for the local community and a 'good quality of experience' for its users (WTO 1993).   It is also hoped that the economic value of the marina will allow for its full benefits to be realised.

Towards sustainable marina development: facing the dangers

Before examining the concept of sustainable marina development, it will be useful to isolate and identify the environmental dangers associated with their construction. In taking account of the dangers, one has also an appreciation of the concept of sustainability.

Habitat loss is found to be one of the most common problems related to coastal constructions. This can be caused by a number of factors, an example being the clearing of mangroves to make way for berths - mangroves being important fish breeding grounds. Pollution can also be a problem, especially in the absence of effective sewage treatment methods and forms of disposal within the confines of the marina. Pollution in the water can cause a build up of algae, which in turn will have adverse effects on the levels of oxygen in the water required to support life forms. Additionally, constructions such as wave attenuators may thwart the natural flushing of a bay or coastal area, which will also have adverse effects on oxygen levels.

Dredging, in order to allow or increase access into a marina, can cause problems in that it can disturb and damage the reefs and plants on which the fish feed and, as a consequence of this, contribute to a decline of fish stocks in the immediate region. Erosion is another problematic factor related to coastal developments. If the siting of a marina changes the tidal regime, then this could cause erosion and have damaging effects on both the natural and man made environment. Visual pollution also needs to be avoided. As in the case of any construction situated along the coast, it is important that a marina blends in with the natural surrounds and is seen to complement the environment.

When designing a marina, it is necessary to take into consideration the factors outlined above in order to achieve sustainable marina development.

Towards an ideal situation - the benefits

Whereas it is important to determine the environmental dangers connected with the construction of marinas, it is just as important to outline their positive features and benefits. Once identified, these benefits (together with the means of achieving them) can be developed and promoted. An example may be given of the siting of a marina on an urban waterfront, such as in Barcelona, which can both improve the appearance of the area and provide environmental and social advantages. This factor is identified by Corrough (*op cit*) who states that

> in many urban waterfront areas marinas have created new habitats, improved water quality and created amenity where there was only decay.

Concentrated shoreline developments such as marinas, providing a wide range of facilities and amenities, may also preserve or even encourage the growth of local natural habitats by decreasing the need for a proliferation of smaller constructions sited along the coast. Larger developments may be easier to control and monitor, and this in turn could lead to greater environmental and sociological benefits.

Marina villages also provide leisure and recreational benefits for the general public, since they act as a 'window on the water' for all and not 'just for the rich with a boat' (Corrough, *op cit*). Economic gains from marinas should, if managed effectively, generate benefits for the local communities. For example employment can be created both during the construction and operation of a marina.

Furthermore, dredging to provide access to the marina will not necessarily have a damaging environmental effect. According to Pearce (1989)

> it may help improve circulation in choked inlets, increase the availability of fish and shell fish, and help to flush and dilute polluted waters.

Towards sustainable marina construction

Once the dangers and benefits of marinas are identified, it is important to determine the means of their construction from a sustainable perspective. Planning in this area should aim to minimise the negative environmental and sociological aspects associated with marinas and their construction, and promote their benefits and advantages.

The Environment

Given the increase in the popularity of boating, Currough (*op cit*) comments that one of the main issues regarding marinas, particularly in the USA is

> the increased need for environmental protection and other natural resources.

As indicated in the introduction, careful consideration needs to be given to the siting of a marina. Some areas may benefit more from the development of a marina than others. Environmental Impact Assessments (EIAs) should provide valuable information in determining the dangers and benefits related to developing proposed sites. Article Three of the E.C. Directive includes:

i.   ecology;
ii.   social and socio-economic matters;
iii.   visual intrusion and landscape;
iv.   geomorphology, existing physical processes, hydraulic and littoral regimes;
v.   air and water quality, pollution;
vi.   noise, odours, vibration, traffic, archaeological, historical and cultural factors;

as bearing considerable importance and significance in relation to an EIA (McKemey 1989).

If an EIA reveals adverse factors concerning the siting of a marina, then the same study must also indicate measures towards mitigating these impacts. Alternatively the choice of another site may be more appropriate. The relationship of a site to other, nearby, harbours and moorings also needs to be considered in both competitive and complementary terms.

Various measures can be adopted to limit adverse environmental and sociological impacts linked to marina development. The use of pile supported breakwaters, to minimise bottom habitat coverage, is one method that might minimise their environmental impact.

Mapping and photographic analysis will prove a valuable form of data acquisition in recording the extent or area of habitats, and thus allow the marina's designers to develop strategies towards their preservation and protection. These surveys can also be used after the construction of the marina for the purpose of monitoring any possible future changes regarding the natural environment and its life forms. In addition, a study of the hydrodynamic conditions, as realised through the use of models designed to monitor wave patterns/intrusions, will be of value in determining the most environmentally sensitive form of marina construction. Interruption of tidal flows (as stated above), transportation or erosion of sediment on beaches, and other changes to littoral zones will have obvious dangers.

It is important to ensure that the quality of water in the marina is good and, if necessary, improved. Plankton and pollutants possibly caused by operational accidents or brought in by storms, need to be dealt with quickly. Flushing the marina regularly is a means of maintaining water quality, although it is desirable that flushing systems will work in conjunction with sewage treatment systems when required. A database listing all the natural resources in the area will also be important towards conducting and sustaining conservational programmes.

The Users

In terms of the needs of the yachting community, considerable importance should be attached to the ease with which they can use their boats, both on entry and within the site. Tidal and sail access is important, as is the positioning of the marina. Pleasant surroundings and good water quality will make the site attractive and possibly encourage longer stays by visitors. This, in turn, should promote greater economic benefits. Road access and other communication links, including bus services to local towns, will contribute to the effective operation of marinas.

Security is one of the prime considerations in determining what constitutes a good marina. Security can be seen both in terms of safeguards against theft, and environmental phenomena such as wind, waves and storms - as aided by effective environmentally sensitive and stable break waters and attenuators. The performance of berthing and mooring structures is also important (Nece & Layton 1989). Research undertaken in Cuba bought to light (mainly through respondents from the USA) indicate that the yachting community desire ease of access to deeper water (natural access channels were favoured) and good lights for entry to and exit from the marina (survey undertaken by the author December 1994). The availability of power and fuel supplies, water for washing and cleaning boats and the availability of materials for boat repairs will all have a bearing on how a particular marina is viewed.

The needs of the local community must also be taken into consideration when designing a marina. The siting of marinas should not 'displace fishermen or other traditional uses without providing alternatives' (Charles 1985). Locals should see the marina as a resource for them, as well as providing the possibility of employment and improving the environment. This can be encouraged through good communications and effective publicity. Water sports for example, and the availability of restaurants and picnic sites, should all prove attractive to those living close to the marina. Given this situation, together with the additional leisure activities on offer, it should be easier to encourage the active participation of the local communities. In Port Moresby (Papua New Guinea) local people helped to fund and build (on a small budget) a much-needed marina. Local labour and materials were used, thus promoting a sense of community pride, value and achievement on the site's completion. In considering marinas and their use, the final report of the OAS Regional Workshop stated that, 'the inclusion of light industrial, commercial, residential, cultural, tourist and any other appropriate user requirements within the area should be a primary goal' (Charles *op cit*). It is quite clear that marinas can have clear and positive benefits for the local community; although the right steps must be taken in order to realise these aims.

A case study of a marine village - Port Solent

The marina village concept has become popular in the UK. Port Solent (Portsmouth) was completed in 1988 on a site of eighty-four acres and has within its confines yachting facilities, chandleries, restaurants, shops, flats, housing and a cinema. Arlington Securities PLC, involved in initiating the Port Solent's marina project, believed throughout that Port Solent Marina would only be viable if the site took the form of a mixed marina/village type of development (Farran *et al* 1989). The design of the marina is quite striking, with the majority of the housing construction affording attractive views of yachts and sea. This site is of particular interest from a planning perspective given that it was built partly on reclaimed land - originally known as Porchester Lake and partly (to the north, where the public utilities are located) on a landfill site. Porchester Lake was filled with rolled chalk obtained from the cliffs close to the marina. In terms of the public utilities site it could be said that the marina exists as a fine example of what can be achieved on contaminated land. The methane gas dispensers located around Port Solent bears testimony to the area's previous use. Many would agree that the existence of a marina, as opposed to a waste site, is a more attractive concept.

The contribution the marina makes to the local area is an important aspect to be considered. Given that Port Solent is a marina village it is hoped that it will be of benefit to a range of different people. Random discussions and interviews undertaken at this marina revealed that many had visited Port Solent on a day trip in order to see the boats. The availability of restaurants and bars provided an extra lure. Others had visited the site during a break from work, business lunches were being conducted and some were receiving sailing instruction. One person remarked that although the marina had a lot of facilities it was not publicised enough, and many people in the Portsmouth area did not know of its existence. Port Solent, however, is advertised in yachting magazines and activities such as bungee jumps, sailing competitions and boat shows are often arranged. Perhaps additional local advertising would increase the number of people visiting the site.

Most of the houses and flats constructed at Port Solent are used as holiday homes, with houses prices (with moorings) starting in the region of £150,000 and two-bedroom flats starting at about £105,000. People living in the flats reported a high level of satisfaction concerning the standard of their accommodation. They were impressed with the facilities provided at the marina and stated that they did use the bars, cinema and restaurants during their evenings out. The shops at the site were tasteful in their presentation and goods-for-sale included souvenirs, clothes, pottery and stationary, although their pricing could be described as 'upmarket'.

Facilities for the yachtsmen were deemed to be excellent, with chandleries and ship engineers readily available. Cameras monitored the site and computerised key cards were essential for entry onto the berths. The water quality in the marina appeared to be good, with strict controls against the dumping of waste. Concerns regarding the contamination of the marina's waters, linked to the land's original use, have been dealt with by the construction of a sheet-piled leachate barrier and drains running along the south side of the marina basin. Periods of free flow of water into the berthing area are undertaken, in order to keep this water's oxygen levels at an optimum amount and thus maintain its freshness.

The majority of the yachts at Port Solent were British owned, although it was reported that quite a few Europeans, particularly Dutch people, used the site in the summer.

Port Solent and the concept of sustainability

In examining Port Solent from the perspective of a sustainable development, a number of favourable points can be identified. It has been built on contaminated land and therefore provides a positive use for an area that has previously been ecologically undermined. The land that had been reclaimed from the sea was polluted marshland and, consequently, the construction of this marina has provided a valuable role in improving the location from both a visual and natural perspective. Local people use the site, although numbers could be increased through greater publicity. It is, however, important to balance the needs of the visitors with the needs of those living there. Additional features such as a swimming pool might make Port Solent more attractive to all. It is quite clear that considerable attention is given to the cleanliness of the marina. Sanctions are taken against those who dump waste and the periods of free flow are designed to maintain the marina's water quality, although these have been restricted to short periods, largely due to pedestrian traffic crossing the locks. It could also be considered that more attention could be given to the enhancement of natural habitats, such as bird life around Port Solent to increase the diversity of shoreline habitats by, for example, providing substrate for fowling communities.

The location of the site is obviously a good one since its sheltered inland position means that it will not suffer too much from wave damage. The design of the marina, with the central berthing area and the housing and chandleries located around it, also means that it is easy to manage and functions well, for day to day usage. Car parking is plentiful and the movement of motor transport does not appear to intrude on waterfront activities. The marina did use fixed berths, although some might argue that those of the floating type would be more sustainable.

Conclusion

The possibility for future marina developments is endless. An increase in the popularity of boating, combined with the comprehensive marina village, provides clear recreational and economic incentives for their construction.

The evident demand for marinas, however, must not be seen to over-ride important environmental and social considerations. EIAs will provide answers regarding the siting of a proposed marina and, given the sensitive nature of coastal zones, attention must be given to the possible effects that marinas could have on tides which, if altered, could cause erosion. It is also important that natural habitats do not suffer as a result of this construction.

The users of the marinas also need to experience and realise their true benefits. Marina designers need to establish what the yachting community requires from a marina and, particularly in the case of a marina village, how it can benefit and attract surrounding communities.

Marina village constructions are already popular in the UK, such as those found in Brighton and Portsmouth. Port Solent (Portsmouth) clearly serves as a prime example of a good and workable marina village for both yachtsmen and non-boat owners. Perhaps more could be achieved with regards to the promotion of this site and the encouragement of local habitats. Nonetheless, it is quite clear that this marina serves as an exemplary model for future developments.

References

Charles, D.O., 1985. The financial aspects and opportunities of improving tourism related facilities in downtown and harbour front areas, *The Final Report OAS Regional Workshop,* St Johns Antigua & Barbuda, General Secretariat, Organisation of American States, Washington D.C., 20006, Economic Secretariat for Economic and Social Affairs, Department of Regional Development.

Corrough, J., 1989. Trends in the planning, design and use of public and private marinas in the United States - forecast for the 90's, *Marinas Design and Operation: Vol. 2,* Proceedings of the International Conference, Southampton, UK, Editors Blain, W.R. & Webber, N.B., UK, 67-84.

Dromgoole, S., Gaskell M. & Grant M. 1989. Planning and legal aspects of marina development, *Marinas: Planning and Feasibility: Vol 1*, Proceedings of the International Conference, Southampton, UK, Editors Blain W.R. & Webber N.B., 17-44.

Farran, R., Farran, N. & Holifield, D.A. 1989. Port Solent - conceptual engineering, *Marinas: Design and Operation: Vol 2,* Proceedings of the International Conference on Marinas, Southampton, UK, Editors Blain W.R. & Webber N.B., 503-516.

McKemey, M.D. 1989. Current requirements and methods for environmental assessment for marina projects, *Marinas: Planning and Feasibility: Vol 1*, Proceedings of the International Conference, Southampton, UK, Editors Blain W.R. & Webber N.B. 165-177.

Nece, R.E. & Layton, J.A. 1989. Mitigating marina environmental impacts through hydraulic design, *Marinas: Planning and Feasibility: Vol 1*, Proceedings of the International Conference, Southampton, UK, Editors Blain W.R. & Webber N.B., 435-450.

Pearce, D., 1989. *Tourist Development*, Second Edition, Longman Group Ltd.

20 Realisation of new urban areas and sustainable development

H.W. de WOLFF

Introduction

Currently urbanisation is again an issue in Dutch spatial policies. The renewed attention has been initiated by the Fourth Report on Physical Planning *(On the road to 2015)*, of which the first concept has been published by central government in 1988[1]. One of the central issues in the report is that new urbanisation areas must be designated. The immediate cause is an increasing demand for space for housing and business activity, in combination with the fact that the existing potential urbanisation areas are unable to meet this demand. In addition, the relationship between urbanisation and mobility (actual use of public or private transport means) is a reason for a reorientation on the urbanisation concept. The high density of the Netherlands and the growing use of private transport, especially in the case of private cars, not only have negative effects on the environment, but also on one of the economic strengths of the country, the Netherlands as a transport and distribution country (with the harbours of Rotterdam and Amsterdam and the airport of Amsterdam as most important 'main-ports'), is threatened.

As a result of the report, new urbanisation areas for the period until 2005 have been designated. Currently the situation for the period after 2005 is in discussion. In several urban areas the realisation of the new urban areas has recently started.

This paper deals with the urbanisation issue in the Netherlands, especially in relation with the possibilities of sustainable development. Firstly, some attention is given to central government policy and the process of designation of the new urbanisation areas. Secondly, aspects of the realisation of the new urban areas will be considered and some remarks made on the land market. Thirdly, the paper focuses on the relationship between sustainable development and the new areas of urbanisation in two themes; the situation of the new urban areas, and the layout of the plans. Finally, some concluding remarks will be made.

Urbanisation in the fourth report on physical planning (VINEX)

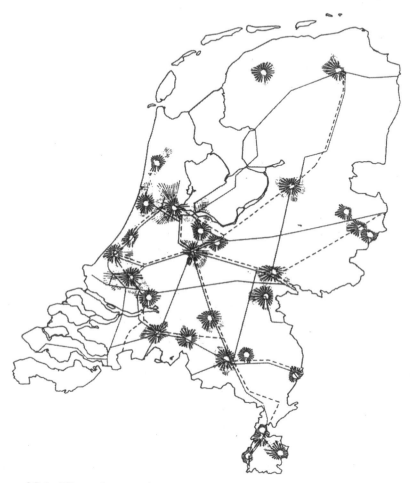

Figure 20.1 The urban regions

In the new urbanisation concept that has been developed in VINEX, two basic assumptions have been of great importance; the reduction of mobility (need to travel), and the protection of rural areas from further development.

Urbanisation is concentrated in, and limited to, 26 urban areas[2], the so called 'urban regions' (stadsgewesten). These urban regions are appointed by central government (see Figure 20.1). Outside these regions, the development of new residential or work areas should only be allowed if such an area is necessary for the local inhabitants.

An urban region is defined as an urban centre and its surrounding nuclei. It can be seen as a functional entity. Most of activities such as recreation, working, shopping and living that are undertaken by the inhabitants of the urban region are concentrated within the urban region.

The new urbanisation areas should be situated within one of the urban regions. However, within an urban region the detailed location of an new urban area is also restricted. For the localisation of new urban functions (residential areas, work areas and amenities) in the urban regions, close proximity to the city centre is an important criteria. Two kind of urbanisation areas are allowed for new development:

1. urbanisation areas that will be located in existing urban areas, close to the inner-city;
2. urbanisation areas that will be located directly connected to existing urban areas.

Existing urban areas

First of all the capacity of existing urban areas should be used. The realisation of new developments in those areas is seen as a important contribution to safeguarding and improving the vitality of the cities. Besides that, the expectation is that mobility will reduce and people will use, instead of their private car, more often public transport facilities. Also, because the distances between different functions are not too large, a growing utilisation of bicycles is expected.

Outside existing urban areas

In some cases, however, there are no more possibilities for urbanisation within the existing urban areas. In those cases, only urbanisation areas that are directly connected to the existing urban areas may be used. Those urbanisation areas, both the residential and the non-residential areas, should be easily reached by adequate public transport facilities and must have a direct connection to the national railway network.

Land Reform and Sustainable Development

Selection of urbanisation areas

A governmental report on physical planning follows a prescribed procedure, the so-called PKB-procedure[3]. At the end of the procedure, the report is approved by Parliament. As a part of this procedure, the provincial, regional and local government involved are consulted, the so-called 'administrative consultation'. The selection of areas for urbanisation has been an important issue in the discussions in this consultation phase.

New housing demand

Shortly after publishing the fourth report on physical planning, results of new research concerning the need of houses indicated that the policy concerning the realisation of new residential areas should be intensified. This new research has been taken into account during the above mentioned administrative consultation. Pressured by these urgent needs, some additional alternative locations have been selected. It was assumed that these new locations could be realised more easily.

Realisation aspects

Unlike the past, the Fourth report on physical planning includes a greater emphasis to the realisation aspects of the physical planning policy. Amongst others, the following aspects are dealt with in the report:

1. concluding of contracts between governments involved concerning realisation of the so-called 'spatial development programme' in the urban regions;
2. stimulation of regional co-operation within the urban regions;
3. public-private co-operation.

Contractual co-operation

In continuation of the formal administrative consultation, central government has been negotiating with the regional governments involved regarding the implementation of the new governmental policy. Central government wanted to reach an agreement with the seven largest urban regions; Amsterdam,

Rotterdam, Den Haag, Utrecht, Arnhem/Nijmegen, Eindoven/Helmond and Enschede/Hengelo, so that realisation of the desired spatial development programme in those regions would be guaranteed.

After a long period of negotiation during the first half of 1995, so-called implementation contracts have been concluded. In these contracts the following agreements have been made:

1.   the local (and regional) authorities involved will realise the spatial development programme that is indicated in the contract. In general such a programme contains the following elements:
   a.   the realisation of the indicated residential areas; the minimal number of houses to be realised is fixed. In contradiction with the past, the largest part of the new houses should be developed in the market sector (at least 70%). A maximum of 30% could be developed in the social sector (by housing associations).
   b.   the realisation of indicated work areas;
   c.   the realisation of indicated infrastructure for public and private transport facilities;
   d.   the decontamination of the soil in the areas to be developed.
2.   central government provides financial support for realisation of the programme:
   a.   a financial contribution related to the expected (negative) net result of the production of land for building in the urban region;
   b.   support for the realisation of the indicated infrastructure;
   c.   support for soil decontamination.
3.   Local, regional and provincial government will implement the so-called 'restrictive policy'. This policy should avoid urbanisation in any other than the appointed areas. The provinces and regions agree to make development of other urbanisation areas, in general, impossible by using their legal powers in the procedure of approving local development plans.

In Table 20.1, the key elements of the implementation contract in the urban region of The Hague ('Haaglanden') is given.

Table 20.1   The implementation contract in the urban region of The Hague

Spatial development programme 1995-2005		
	Residential areas	Work area
• Inside existing urban areas		
	9000 houses	pro memoria
• New urban areas		
Wateringen	6500 houses	57 ha
Ypenburg	9000 houses	57 ha
Nootdorp Stuno	1500 houses	
Leizo	6800 houses	40 ha
Zoetermeer-Oost	6000 houses	17 ha
Pijnacker Zuid	1700 houses	10 ha
Pijnacker Delfgauw	2000 houses	50 ha
Financial support from central government		
• Production of building land		F 263 M (125M ECU)
• Realisation of infrastructure		F 300 M (143M ECU)
• Cleaning up the soil		F 10 M (4.8 M ECU)

This method of implementation of central government policy is rather new in the Netherlands. Normally, instruments of administrative law (e.g. approval of local plans) as well as granting schemes were used for policy implementation by central government. Using this new approach, the role of central government has changed from a more passive one towards a more active and developing one. Instead of focusing on the resources, central government now focuses on the results.

Regional co-operation

For successful implementation of the new planning policy, obligatory regional co-operation is an important precondition because:

1. a large amount of land is developed for the market (housing in the market-sector and other commercial use). Therefore, competition between communities must be avoided, because such a competition may have a negative effect on the land prices that can be obtained;
2. the financial contribution from central government for the production of building land is based on the deficit of all the production of building land in the urbanisation areas in the region. Therefore, 'lucrative' urbanisation areas must pay for urbanisation areas with a negative financial result. This can only be realised with obligatory regional co-operation, because very often the 'lucrative' urbanisation areas are situated within the boundaries of another municipality rather than the urbanisation areas with a negative financial result (e.g. inner-city areas).

For facilitating regional co-operation, a new act has been made by the government that came in force July 1994, the 'Enabling legislation for regional co-operation' (Kaderwet bestuur in verandering). The range of the act is limited to the seven largest urban regions. The act obligates local authorities in the urban regions involved, to establish an administrative body for regional co-operation. This new body should, at least, receive some indicated powers. One of these is the obligation to make a *regional plan*. This plan will play a role in the approval process of local development plans.

The administrative body can also constitute a *regional land policy*. In this policy, directions can be given to local authorities (e.g. in the field of the acquisition of land, the disposal of the developed building land (prices, the phasing of the disposal), and the division of costs related to land development among the local authorities within the region). Besides these directions, the administrative body also has the power to realise some land development projects itself, by buying land, realising infrastructure and public services, and finally selling the building land to housing associations and private developers.

241

Co-operation between the public and private sector

Central government stresses that the realisation of the new urban areas cannot be done by local governments on its own. Co-operation between the public and the private sector (the so-called public private partnership) will be necessary for realising the intended spatial developments.

Land market

For the realisation of the spatial development programs in the urban regions, the availability of land is a key factor. Traditionally the land market for urbanisation purposes in the Netherlands was dominated by local authorities. This was due to the need for a large amount houses to be realised in the social sector, within those urban areas. However, in this field great changes have occurred. In 1989, central government presented a report on the housing policy in the nineties (*Volkshuisvesting in de jaren negentig*); a central point in this report is the liberalisation of the housing market and the withdrawal of governmental interference in this market. The possibilities for grant aid will, therefore, be greatly reduced. As a consequence of this, housing in the new urban areas will, for the larger part, be realised in the so-called market sector.

The effect of this change of policy on the new urban areas, especially the urban extension areas, has been enormous. Traditionally, private developers and building companies were very cautious about buying land in such areas, because they were not allowed to realise social housing. When a large area was used for social housing, the local government could use its dominated position to discourage the private developers, for instance by slowing down the development of the plan for the areas owned by the developers. Also, the localisation of social houses to be developed on the land owned by the private developers or building companies was a successful strategy followed by local governments.

However, the situation has now changed enormously. First of all, because of the relatively public negotiations about the implementation contracts, the urbanisation areas to be developed are indicated very clearly. For private developers, there is only a small risk of buying land that will not be developed in the near future. Secondly, because of the large share of market sector housing in the new urban areas, the above mentioned risks for private developers of a non-co-operative local government have been reduced. Thirdly, because of the already mentioned restrictive policy, only a limited number of new urban areas will be developed in the future; for gaining an opportunity to participate in the purchase of land will be a prerequisite. The result of the new situation is that private developers and building companies have become quite active on the land market.

These changes have several effects:

1. land prices have risen enormously. In some areas the prices that are paid for agricultural land per square metre have risen from $f7=$ per m^2 (3.3 ECU) to $f40=$ per m^2 (19 ECU);
2. a substantial part of the land in new urban areas is owned by private developers and building companies. Generally, it concerns the most easy-to-develop parts of the urbanisation area[4]. Local authorities have no alternative but to cooperate with the private developers in realising the new urban areas.

As a reaction to the declining role of government on the land market, an enactment has been made to accomplish a pre-emption right for local governments in new urban areas. This pre-emption right should have the following effects:

1. to subdue the rise of the prices of agricultural land to be developed as a part of an urbanisation area. Because of the pre-emption right, local government gets the first right to negotiate. However, when the parties involved do not agree and still want to reach an agreement, the price can be set by experts using the criteria from the Expropriation Act, which means that the price is derived from the new value of the land, based on the future permitted land use.
2. to, once again, give the central position of local government to directing land development.

However, the enactment is late. It came into force in 1996 and, for the new urbanisation areas, the 'damage' has been done.

Sustainable development

In the preceding paragraphs, the context of the realisation of the new urban areas in the Netherlands has been indicated. At first sight, the combination of land reform, a more general governmental intervention in the land market, and sustainable development may seem a strange one. Often the first association of sustainable development is an economic one. Our current way of living and the 'necessity' of economic growth cannot possibly last for ever. Incredible damage to the environment has already been done; future generations will face major problems. Sustainable development has to be the process used to stop this cycle. An adapted way of consumption, using less energy and less scarce resources, is an important condition for sustainable development. However, in the free market economy, where individuality is a key achievement, the changing of consumption patterns is a very long-lasting process.

Land, however, is an important economic factor and in many countries the land-market is traditionally the market where governmental influence is rather important. This influence might be used for realising sustainable development. In the Netherlands, as part of the new urbanisation policy, elements of sustainable development are integrated. Two aspects will be considered; the situation of locations for new urban areas and, the potential for using the principles of 'durable buildings'.

Space for new urban areas

Space is scarce, especially in the Netherlands. During the last few decades, an enormous amount of rural land has been used for urbanisation purposes and for the realisation of new infrastructural projects. There will always be a tension between urbanisation and sustainable development: urbanisation in itself poses a claim on scarce land and this cannot be maintained for ever. However, the new Dutch physical planning policy of the Central Government takes this threat into consideration. Avoiding uncontrolled development in rural areas is, as has been mentioned, one of the reasons for limiting the number of new urban areas, next to reducing mobility.

More intensive use of existing urban areas

The more efficient use of existing urban areas should be a major item in governmental policy, from the viewpoint of restricting the claim on scarce land. There are a number of possibilities for reaching this objective. Some examples include:

1. in cities in the Netherlands old industrial areas can often be found which are no longer intensively used;
2. because of recent changes in the market for office space in many cities throughout the Netherlands, the number of unused office space is increasing. More and more office activity is concentrated close to railway stations or highways;
3. due to improving living standards, many people live in houses with much more floor space than in the past. However, elderly people do not move when their children have left the house, even though they need less room space.

Different governmental projects have been set up to address these points. The redevelopment of old areas has been, and still is, an important goal of **urban renewal** in the Netherlands. The government has played an active role in trying to buy the land, redeveloping it, etc. However, some problems occur nowadays. Firstly, the amount of money Central Government provides for urban renewal activities has been reduced. Secondly, the soil of many old industrial areas is polluted; the possibilities for financing the soil decontamination by the government are, due to new regulations, limited. These new regulations insist that initially, the polluter or the present owner has to pay for the decontamination. This can easily lead to a stalemate, which results in no activity at all.

Some projects have started without direct governmental intervention through acquiring the property. In these projects, an attempt is made to convince the owner of a building to change the use of the building. Already some good examples of changing offices into apartments have been realised. However, the possibilities for the government to realise more intensive use of existing buildings without direct intervention are limited. People cannot, without some limited exceptions, be forced to use their buildings more intensively. Local government seldom has enough money to establish a convincing scheme of grants. Recently, the implementation of fiscal incentives is under discussion.

Nevertheless, the more intensive use of existing buildings and existing areas in the cities is of great importance for reducing the amount of new land that is needed for urbanisation. Besides that, it can also have positive effects on mobility and, therefore, on the use of energy for transportation.

Urban extension areas

The new urban extension areas that have been appointed are mostly adjacent to existing urban areas. However, it is still an open question as to whether the aimed reduction of mobility will be realised. Not all new urban areas will have direct access to the national railway network. Furthermore, the Dutch railway company is rather reserved with regard to allowing the opening of new stations. Also the accessibility of the new urban areas by other means of public transport can cause problems. The existing grant scheme only allows new facilities to be realised when a certain amount of use (and therefore profit) can be realised. However, during the first years of the development of a new urban area, it is unlikely that this will be realised. Experiences indicates that when people have got used to a situation where public transport is not present, it is very difficult to change their behaviour when good facilities are realised.

Some disadvantages

Although some implementation problems occur, recent Dutch urbanisation policy seems to contribute to sustainable development. However, there are also some disadvantages in relation to sustainable development of the policy of limiting urban development to a certain number of urban regions and specified locations close to the city centre. These include:

1. very often the locations in or close to the city have previously been used by industrial activities;
2. frequently soil pollution is discovered in these areas. The decontamination of the soil is rather expensive;

3. furthermore the intensive use of urban areas causes negative side-effects: living next to a railway line or a highway, for instance, gives rise to noise pollution and can be dangerous in case of an accident;

4. the level of air-pollution in densely urbanised areas can become rather high;

5. partly because of the above mentioned problems, more and more people want, for recreational purposes, to leave the city. The recreational possibilities within the urban region often are intensively used. More and more people travel long distances, often by plane, for taking some days of. This causes more and more mobility and therefore use of energy.

Another disadvantage is a more principal one. In general a development can be notified that the relation between production and consumption of economic goods is changing. The scale is getting bigger and bigger. For instance, the production of food does not any more take place within the region where the consumers live. This economic specialisation causes more and more mobility (green beans for instance are transported by plane from Egypt to the Netherlands, flowers are transported from South America towards Europe, etc.). There is no sense and therefore no responsibility by the consumers with regard to the circumstances under which production takes place. The strict division between urban and rural areas, proposed in governmental policy, seems to confirm this developments.

Layout of and buildings in the new areas

Not only the location of new urban areas, also the design or the lay out of the areas and the buildings to be realised must be in line with principles of sustainable development. In the implementation contract, the governments involved have agreed that the principles of 'durable building' will be used by the realisation of the new urban areas. This means a careful selection of building materials, energy saving constructions, etc.

If these principles can be realised, however, largely depends on the governmental position on the land market. Many aspects can not be forced upon developers by administrative law. So if private developers owns the land, they cannot be forced to be committed to all principles of durable building, for example.

The Dutch Housing Act used to allow local government to make their own local bye law with their own building regulations. However, recently this act has been amended. Laying down building regulations has been centralised. Only the central government regulations as well as the local development plan might be taken into consideration when taking a decision on an application for a building permit. In the regulations of central government, no attention has been paid yet to environmental aspects (e.g. durable building).

In contrast with public regulations, in private contracts aspects regarding durable building can be taken into account. So if the local government owns the land, some demands in this field can be regulated in the deed of conveyance between a private developer or a housing corporation and local government. Also when a public-private partnership is established, to a certain extent agreements on durable building can be made.

For most of the new urban areas public private partnerships will be of great importance. Generally, private developers do not seem to oppose to the general principle of the durable building concept. So aspects of a durable layout of and durable buildings in the areas might be realised.

However, still some problems rest in this field. Recent research indicates that the single-family dwelling, or more specific: the 'house with a garden' still is the favourite one. That means that trying to realise high densities in the new urban areas can cause problems. High density is not only necessary for protecting the open spaces, it is also a prerequisite for the possibilities of good and frequent public transport facilities. The accessibility of an urbanisation area by car is also greatly approved by many people. However, this may cause a less frequent use of public transport facilities.

Of course private developers will listen carefully to such signals of the market. So realising a higher density will cost local government an enormous power of conviction because it seems to make the product of the developer less profitable.

A consequent regional policy is of great importance, so that competition between different areas will not occur. The new possibilities for regional co-operation might facilitate such a regional policy. However, until now the regional bodies hesitate to use their powers.

Central government has launched a scheme to stimulate durable building. Part of the schedule is a granting scheme. Grants can be obtained for the use of durable materials, for special technical facilities for saving energy and water, etc. Also the realisation of good examples is stimulated.

Conclusion

Major changes have occurred in the field of realisation of new urban areas in the Netherlands. The concentration of urbanisation on a selected number of locations, the realisation of good accessibility by public transport facilities are part of this change in governmental policy.

The policy change is complicated by the declining dominant role of local government on the land market. This seems to make realisation of the concept of sustainable development more complicated. More important however seems the fact that the final 'consumers' of the new urban areas must be convinced that a sustainable location is preferable to a classic one. Winsemius, a former Dutch Secretary of State on the Environment, uses the term of 'self-regulation'. When new policy-goals should be realised, there are three different ways for the government to implement the policy:

- direct regulation: the government forbids certain activities or obligates people to change their behaviour;
- indirect regulation: the government uses financial stimuli so that the new behaviour is rational in an economic way, as a consequence of this the behaviour is changed;
- self regulation: people changes their behaviour on their own initiative, because they believe the new behaviour is the right thing to do.

The most important task for central government therefore will be to make sustainable development, also in the field of urbanisation, the normal kind of development.

Notes

1. The so called *Vierde nota over de ruimtelijke ordening* (VINO). The draft version has never got the formal approvement of Parliament, because of a

governmental crisis in 1989. The next government introduced an extended version of the report, the *Vierde nota over de ruimtelijke ordening Extra* (VINEX), in which some elements are added. The VINO pointed strongly on consolidating the strong points of the country. In VINEX more attention is paid to weaker functions and inhabitants. Also more attention is paid to aspects of realisation of the new policy.

2. Of course much political struggle and lobbying has been done. The final number of 26 urban regions can therefore rather be seen as a compromise, than as a purely rational choice.
3. PKB is a Planologische kernbeslissing, a report that contains policy decisions from central government concerning planning matters.
4. For instance, in the new urban area 'Wateringse Veld' in the urban region of The Hague, the larger part of the meadow-land has been bought by a private developer. However, the land on which warehouses are build has not been bought.

21 The railway town: a case study in sustainable urban development

R. W. DIXON-GOUGH

Introduction

In all parts of Britain, and indeed in any part of the World, the railway has exerted a profound influence upon both the landscape and the social and economic development of the population and the land. The resulting changes are, in many respects, limited to areas 'touched' by the effects of the railways. Thus, the greater the density of the railway network, the greater will be its influences upon the land and the population. The effects in Britain, with the very dense railway network developed during the nineteenth century, are therefore much greater than would be experienced in many other European countries. Barber 1970, states that:

> The activities of the Railway Companies in the nineteenth century had important social and topographic repercussions on both the urban and rural environment.

The railways not only had an important effect upon the environment but also upon the lifestyles of the population. Trevelyan (1979) however points out that the main effects of the railway lay not in changing people's lifestyles or even the environment, but in accelerating those changes. With respect to lifestyles he considers, for example, the recreational activities of the population:

> Even before the age of the railway, Londoners had swarmed on the pier at Brighton and darkened the sands of Margate with their multitude. Now the whole coast of England and Wales was opened out to 'tripper' and 'lodgers', by steam locomotives and by the increased earnings and savings of all ranks.

In a similar manner, Lloyd (1984) points out that:

> Most of the major Victorian towns were already well-established before the railway age, apart from a few which were primarily ports or resorts, and some which owe their existence to the railway's operation.

As in the case of recreational activities, the process of industrialisation had commenced in certain regions by the beginning of the nineteenth century. It was, however, located in the proximity of accessible reserves of coal and iron and thus, an alien activity in the southern counties. In the early part of the nineteenth century there appeared to be little likelihood of industrial growth taking place away from traditional locations on the coalfields of the Midlands, northern England and Scotland (Turton 1969). In the regions away from those industrial zones, the main centres of population were the market towns and these had little or no interest in manufacturing industries beyond the range of domestic crafts. Some local industries, such as quarrying and textiles existed across the country but the majority of these were carried out on a relatively low scale of intensity and mainly to satisfy local needs.

However, with the rapid expansion of the railway network during the early years of the nineteenth century, the railway worker became a familiar part of the labour forces in both urban and rural areas. In just ten years, between 1841 and 1851, the number of men employed in operating the railway network in Britain increased from 2,000 to more than 25,000. In addition to those directly employed to operate the railways, there was a supporting force of maintenance workers and others engaged in the actual construction of locomotives, carriages and wagons. The railway companies were quick to recognise the advantages of establishing these workshops at nodal points on the individual networks. Some workshops were opened in towns such as Derby, Manchester and Glasgow, which were already established as industrial centres. Others, however, were located at strategic railway junctions in areas remote from the coalfields and the manufacturing regions. These provided the basis for the growth of a distinctive category of Victorian industrial community which rapidly became popularly identified as the 'Railway Town' (Turton op cit).

Once the railways reached the rural areas of the south of England, the development of such manufacturing centres was rapid. By 1850, expanding and thriving centres of railway engineering had been established in Swindon, Wolverton and Ashford. The subsequent industrial growth and development of these towns can be directly related to the expansion and contraction of the railway network as a whole. Furthermore, the history of each community is closely associated with that of one of the railway companies, which existed prior to the railway grouping of 1923.

Many of these towns were developed in rural areas, often some distance from the nearest market town, and with no 'pool of skills' available in the locality. Thus, the initial labour forces of the railway workshops were skilled workers attracted from other engineering and manufacturing areas. Invariably, the local housing was insufficient to accommodate the incoming labour forces and the railway companies themselves financed the building of dwellings. This combination of railway world and railway housing provided the core of the embryonic railway townships. The Railway Town had arrived.

The Railway Town

The railways, or railway companies, did not generally make towns. They did, however, contribute considerably to the founding of many towns and played a part in generally concentrating populations (Perkin 1970). The contribution of the railway companies in the development of such towns can be illustrated by giving two examples; recreational and industrial. Holidays by the coast had become a regular part of life to the lower middle class by the early part of the nineteenth century, and even to large sections of the working class, particularly in the north. This exodus to the sea was greatly assisted by the railways and certain coastal towns with rail links expanded rapidly to meet the recreational demands of these people. By 1876 Blackpool, a coastal town in the north west of England - close to the industrial regions of Lancashire and Cheshire, had grown to the size and status of a Borough. The annual holiday of the Lancashire and Cheshire artisans also supported the development of towns such Llandudno and Rhyl (along the North Wales coast) and the Isle of Man as well. Cornwall, in the west of England, was already the holiday resort of the well-to-do at Easter and the masses in August, whilst in the summer the lodging-houses in Keswick and Windermere, and the farms of the Lake District were thronged with family parties (Trevelyan 1979). These were very different types of

Railway Towns, not made by the railway companies, but greatly affected by the introduction of the railway and mass travel.

The second type of Railway Town was the type that owed its existence to the railway's operation. Of these, the most remarkable were Swindon and Crewe, two towns effectively developed by railway companies to serve the needs of a locomotive works built in rural areas. The establishment of a new locomotive works and the associated infrastructure of the dwellings necessary for the workers might seem to be an enormous undertaking on the part of the railway companies. It should, however, be considered that during the nineteenth century, more than £1,000,000 (an eighth of railway capital) had been expended in urban demolition and reconstruction - making the development of a rural site appear very attractive and relatively cheap. The Great Western Railway adopted this approach by setting up its works, where most of its locomotives and rolling stock were eventually made, in open country near the very small market town of Swindon. Brunel and Wyatt designed for the railway company a model community of worker's houses, with social and community buildings, most of which have been recently restored (Lloyd 1984). It was:

> small, modest and laid out with ingenuity (Pevsner, 1963).

It is this second category of railway town that is to be the subject of this chapter. The sustained development of Swindon will be examined, through its evolution, from a small market town in a rural backwater to becoming the home of one of the major locomotive works in Britain. In turn, those locomotive works have completely closed and new industries and economic structures have had to be developed.

The evolution of Swindon

The evolution of Swindon as a major commercial centre has been a gradual process of evolution brought about by a series of sustainable developments that may be conveniently be classified into four separate stages:

1. the development of New Swindon as a Railway Village, 1840-1851;
2. the expansion of the Railway Workshops and the encouragement of private industrial growth by the Railway Company, 1852-1918;

3. the recognition by Swindon Borough Council of a need to sustain development by diversifying employment opportunities, 1919-1945;
4. the contraction of the Railway Workshops and the development of Industrial Estates, 1946 - date.

The first period of sustainable development - the Railway Town

This period is concerned with the development of New Swindon and the integration of the old and new towns. Old Swindon was a small market town set in an agricultural county and the following description is very appropriate:

> The town is situated on the summit of a hill of considerable eminence, which commands a delightful view of parts of Berkshire and Gloucestershire. The principal streets are wide and contain many good houses. No particular manufacture is carried on in the town but it is the residence of many persons of independent fortune. Extensive quarries are wrought in the neighbourhood, which, together with agricultural pursuits, affords employment to the greater of the working population of the town. The inhabitants are abundantly supplied with pure water from springs. There are 325 houses in the town and the population at the 1831 census was 1,742 (Anon. 1843).

Any attempt to develop this town as the location of the new locomotive works could not succeed, partly due to the its' location on a hill, but also because the existing infrastructure would be quite unsuited to the demands that would be placed upon it by the new railway works. Without the railway village, the locomotive works could not have operated. The old market town was a mile away from the railway and quite incapable of either housing or lodging the workers on the scale required (Hudson 1968).

A creation of the railway town, or New Swindon, was entirely the achievement of the Great Western Railway and the town is a product of the Industrial Revolution. A variety of reasons led to the development of Swindon as a railway town. Those technical considerations include:

1. changes in the gradient of the line between London and Bristol necessitating, in those days, a change in locomotive type;
2. Swindon was to be the location of a junction between the Great Western Railway and proposed Cheltenham and Great Western Union Railway;

3. the location of the Wilts and Berks Canal giving direct connection with the Somerset coalfields permitting easy transportation of coal, and an adequate supply of water for the locomotive works;
4. a flat 'greenfield' site;
5. Swindon effectively divided the route from London to Bristol into three, roughly equal, sections; London - Reading, Reading - Swindon, and Swindon - Bristol.

A recommendation was put to the Directors of the Company who decided:

> to provide an Engine Establishment at Swindon, commensurate with the wants of the Company, where a change of engines may be advantageously made, and the trains stopped for the purpose of passengers taking refreshment. The Establishment there would also comprehend the large repairing shops for the Locomotive Department, and the circumstances rendered it necessary to arrange for the building of cottages, etc., for the residences of many persons employed in the service of the Company (MacDermot 1964).

The line reached Swindon in December 1840 and by June 1841 the entire line from London to Bristol was in use. Plans were quickly made to build 300 cottages and land was bought to allow work to commence on the construction of the locomotive works and the Railway Village. The development of Swindon as a railway town had commenced.

The cottages of the railway town were laid out in a symmetrical, grid-iron pattern, administered by a specially constituted committee drawn from the GWR Board and by 1853, about 240 houses had been provided. Wide streets separated the terraced cottages so that the living conditions of the workers were light and airy for the period. They were between 10 and 15 years ahead of the main body of law regarding the construction and use of 'labouring-class dwellings' (Vaughan 1977). Attention was also paid to the spiritual, educational and physical welfare of the companies employees with a substantial church (1844), schools (1845) and a park - all endowed or aided financially by the company.

New Swindon, as a town had several remarkable characteristics. For example, unlike many towns, which expanded rapidly during this period, it had a single focus of settlement (the Great Western Railway Works) and not, as in other industrial towns, several foci formed by scattered groups of factories. This resulted in a compact town, built essentially around the Railway Works. Furthermore, the growth of the railway town was unrelated to any existing road systems radiating outwards from the centre of the town, as in the case of the majority of developing towns. There was, therefore, no temptation to ribbon development. This factor further emphasised the compact nature of New Swindon (Wells 1950). These factors resulting in the Board of Trade (1908) reporting that:

> Swindon is one of those towns which are created rather than developed, owing to their selection by a company or some similar industrial organisation as the scene of its operations.

The second period of sustained development - expansion of the railway workshops

By 1851, 92% of employed persons in Swindon worked in the railway workshops. Between 1851 and 1911 the population of Swindon rose from 4,876 to 50,751 (Table 21.1.). It had drawn in many immigrants from the adjacent rural counties of Wiltshire, Gloucestershire and Somerset, particularly from Bristol and the declining cloth manufacturing districts, but many more from London, the north of England, Scotland and Wales. The task of housing these people, providing them with adequate amenities and welding them into a community was a formidable one and only the paternalistic and immense prestige of the Great Western Railway made it possible to carry it through. New Swindon is, today, much as it appeared in 1850. It is:

> one of the few planned Victorian estates, small and modest, and architecturally as orderly as is the design of its streets (Pevsner 1963).

In many respects, the people who were employed by the railway and lived in the New Town were a privileged group compared with the working class of Old Swindon and the surrounding area. They earned more than most other Wiltshire people, they enjoyed remarkably good amenities which the Company either provided or supported and they could reckon that their employment was reasonably secure, so long as they kept good health and on the right side of the foreman. There were considerable disadvantages in working in a one-industry town, and the Swindon railway workers buying their own houses had much to lose. By 1870 Swindon was, in a very real sense, the Great Western and if a man, or his wife, fell out with the Great Western there was very little for him to do but leave the town. The endless terraces of small houses stretching away from the railway are a monument to fear, and paternalism, as well as to thrift and engineering skills (Hudson 1968).

Table 21.1 The population growth of Swindon.

Year	New Swindon	Old Swindon	Total
1821		1 580	1 580
1841		2 549	2 459
1851	2 300	2 576	4 876
1861	4 167	2 689	6 856
1871	7 628	4 092	11 720
1881	15 086	4 818	19 904
1891	27 295	5 544	32 839
1911			50 751
1931			62 418
1945			67 000
1975			130 000
2000*			200 000

Source: Census records. Projected figure (Thamesdown Borough Council, 1985).

The second half of the nineteenth century became a period of sustained development for the town of Swindon and this can be seen in the general increase in population (Table 21.1). The railway works became increasingly important as the company expanded territorially: the locomotive

258

building at the works of constituent companies, located at Newton Abbot, Saltney (Chester) and Worcester was transferred to Swindon later in the nineteenth century. In 1869, further growth arose because of the concentration of coach building at Swindon - the proposed site at Oxford being rejected by the Borough. Swindon Works also produced its own steel plate, had a central laundry and depots for clothing and kitchen equipment.

The company also provided services including a private health service in 1847, the Medical Fund Society eventually becoming extremely comprehensive and included hospital facilities at Swindon in 1872.

What should be recognised is that the Great Western Railway, together with many other large employers, was a very 'paternalistic' organisation, although this factor could be open to interpretation! An element of this attitude was directed towards the town, as well as its employees, and this was manifested through encouraging the growth of private companies already established in the area. Thus, the Castle Foundry (1855) produced bridges and railings for the Great Western Railway as well as milk churns and dairy appliances for the surrounding area. A brick works, opened in 1871, supplied the railway's needs, as did two clothing works, which opened in 1876 and 1899. A clothing manufacturer started business in 1912 and a tobacco factory in 1915.

The third period of sustainable development - industrial diversification

The first opportunity for industrial diversification, and large-scale female employment, in Swindon came with the transfer from London of a firm of uniform clothing manufacturers. The firm acquired a contract from the GWR for the supply of uniforms. The factory came to Swindon in view of the town's status as the central stores of the GWR, the offer of a site adjacent to the station and preferential traffic rates. It was also attracted by availability of female labour in an area, which at that time, had no other female employing industries. Other clothing factories followed in 1901 and 1918 together with a tobacco factory in 1915 and a gramophone works in 1919. Women predominated in the labour forces of these industries. By 1921, 35% of the manufacturing workers living in the town were employed in activities other than railway engineering.

The paternalistic attitude of the Great Western Railway was still very much in evidence during this period. After the First World War, the Railway was approached by its employees to help meet their trading needs.

Railwaymen often had to re-locate and could not, therefore, buy their own houses, and local authorities - at that date - were giving priority to those who had lived in the localities before the war. The company advanced £1,400,000 in direct loans enabling over 3,000 employees to buy their houses, and also formed Public Utility Housing Societies of whose capital nearly £750,000 were provided by the Company (Collins 1971).

Table 21.2 Numbers employed by the railway company (the Great Western Railway and British Railways)

Year	Number Employed	Percentage of Working Population
1843	417	28%
1847	1 800	91%
1877	13 000	80%
1921	14 000	53%
1931	12 000	45%
1937	10 000	42%
1962	8 000	26%
1963	6 800	21%
1967	3 000	10%
1973	2 200	6%
1980	3 800	9%

The successful history of sustained economic development experienced by the railway towns during the nineteenth century reflects the status of the railways as one of the leading growth industries of Victorian Britain. By 1921, the railway employment at Swindon reached its maximum, and the total employed now numbered about 14,000 (almost 53% of the working population) (Peck 1983).

The gradual contraction in the scale of railway operations, initiated in the 1920s and accelerated after the Second World War, has resulted in a steady decline in the production capacity of the railway workshops and a corresponding reduction in their size of the labour market. By 1930, the percentage of the working population employed in the Railway Works had declined to 45% and was reduced throughout the 1930s, illustrated in Table 21.2. These decreases, coupled with the introduction of other industries, led to the emergence of more balanced economic structures in the railway towns. Swindon was no exception to this. During the 1930s, Swindon Borough Council, wishing to avoid the social and economic problems associated with an ill-balanced industrial structure, made great efforts to attract other industries to the town. These largely complimented the trades and skills used in the Railway Works and were invaluable in maintaining employment levels through this period.

The figures given in Table 21.2 indicate the degree to which the numbers employed in the Railway Workshops contracted. The main period of contraction commenced in the early 1960s but even before then, there were indications that other industries must be attracted to Swindon to provide alternative employment opportunities. The major employers prior to the Second World War were the Great Western Railway Works, W.D. and W.O. Wills (tobacco), Garrard (Engineering), and the Cellular Clothing Company.

The fourth period of sustainable development - the contraction of the railway workshops and the development of industrial estates

In 1944, the Planning Department of Swindon Borough Council recognised the need to "refit the one-industry town":

> The only sound insurance against unemployment is to have more than one industry.........What is required is the diversification of industry within every area (Swindon Borough Council 1944).

This memorandum became a policy document and, coupled with the Government's policy concerning New Towns, led to housing developments aimed at resettling Londoners. In parallel with the programme of resettlement came the industrial diversification programmes, which commenced with the development of a Trading Estate in 1956 and which, by

1960, provided employment for 5,200. Some of the first companies to occupy this Estate transferred from London, bringing their workers and families with them, yet at the same time expanding and offering employment opportunities to local people. In 1957, a second Trading Estate was opened, this one being largely occupied by engineering and metal working companies, both trades based upon the redundant skills of the Railway Works. By 1964, a third Trading Estate had been developed (Cullwick 1975). This had the effect of moving the focus of employment from the centre of the town to foci around the perimeter, close to the newly built housing estates.

Table 21.3 The percentage of workers in Swindon employed in the four main sectors

Occupation	1945	1971	1981	1991
Agriculture	5.4	2.3	2.1	1.3
Construction	7.8	5.6	5.8	6.4
Manufacturing	68.3	45.4	26.9	17.2
Services	18.5	47.3	65.2	75.1

In the period up to 1968, over 100 major employers had re-located or located in Swindon. Employment in the Railway Works had fallen to 3,000 and it was now the third largest employer after Plessey and Pressed Steel. During the 1970s, six further industrial estates were developed and a further 42 enterprises re-located to Swindon. More recently, there has been an increased move to re-locate electronic, computer and other high-technology companies to Swindon taking advantage, this time not of the rail links but of the 'M4 Corridor'. During this period, service industries began to replace manufacturing as the main source of employment. The changes in the labour force are listed in Table 21.3. This emphasises the development within the town that has been sustained since 1945, and of the nature of the employment. Major factors in the growth of this sector of the service have been identified by Thamesdown Borough Council (TBC, 1991) as improved communications (prompted notably by the motorway and the development of the High Speed Train).

This period has also witnessed a considerable degree of diversification. By making the town less reliant upon a single industry, it will inevitably place the borough in a position to encourage a sustainable level of industrial growth and employment, since any effects of world recession upon industrial growth will be spread. In addition to the service sector, there has been an increase in the number of light industrial concerns locating within the area, particularly those involved in medical engineering and the high technology sector. Most of those companies have been located in purpose-built premises on one of the 25 industrial estates, each of which has easy access to both the local labour force and the major communication networks.

Finally in 1986, the railway workshops closed. During this latter period of sustainable development, over 8,000 jobs had been lost from the workshop but the redundant workforce found work quickly in the new industries of the town and district (Peck *op cit*). It is somewhat ironic that some of the historical buildings of the former railway works have been converted into a retail centre. Nowadays, those old buildings echo to the sound of thousands of voices of women, children and men seeking out the best bargains and eating fast-food, where once the locomotives of the Great Western Railway were built and repaired.

Conclusion

The development of Swindon as a railway town and of its subsequent development as a regional and industrial centre could only take place providing that a programme of sustainable development was maintained. Swindon is perhaps fortunate in that it was a 'one-industry' town but that that industry made efforts to encourage the diversification of industry in Swindon. It is also important that as the influence of the Railway Company waned through the recession of the 1930s, the enlightened actions of Swindon Borough Council and its successor, Thamesdown Borough Council, took on the challenge with the development of important policies. Finally, the current state of the industry within Swindon owes much to its proximity to the M4 motorway. As the railway declined in importance, so the motorway became the major influence upon sustainable development in

Swindon. This process can best be summed up by the Corporate Strategy for Economic Development of the Borough (TBC 1995)

> To work in partnership with all relevant agencies to develop Swindon as a balanced sustainable regional centre and to ensure that the benefits that accrue from this development are enjoyed by all sections of the Borough's population, safeguard the environment and provide the necessary infrastructure.

References

Anon. (1843) *The Parliamentary Gazetteer of England and Wales 1840-1843.* A. Fullerton and Co.

Barber, B. (1970) The concept of the railway town and the growth of Darlington, 1801-1911: A note. *Transport History*, **3(3)**, 283-292.

Board of Trade (1908) *Cost of Living of the Working Classes.* (Cd 3864), CVII, 319.

Cattell, J. & Falconer, K. (1995) *Swindon: the Legacy of a Railway Town.* HMSO, London.

Collins, H. (1974) *Britain's Railways - An Industrial History.* David and Charles.

Cullwick, C. (1975) *Industrial and Commercial Expansion of Swindon, 1945-1975.* Unpublished Dissertation, University of Hull.

Hudson, K. (1968) The early years of the railway community in Swindon. *Transport History*, **1(2)**, 130-152.

Kellett, J. (1969) *Railways in Victorian Cities.* Routledge and Kegan Paul.

Lloyd, D. (1984) *The Making of English Towns. A Vista of 2000 Years.* Victor Gollancz Ltd.

MacDermot, E. (1927) *History of the Great Western Railway. Volume 1, 1833-1863. Revised by Clinker, C. in 1964.* Ian Allan.

Peck, A. (1983) *The Great Western at Swindon Works.* Oxford Publishing Company.

Perkins, H. (1970) *The Age of the Railway.* David and Charles.

Pevsner, N. (1963) *The Buildings of England: Wiltshire.* Penguin.

Silto, J. (1981) *A Swindon History, 1840-1901.* Library and Museum Service, Wiltshire.

Swindon Borough Council (1944) *Memorandum on Potentialities for Industrial Development.* Planning Report. Civic Offices, Swindon.

Thamesdown Borough Council (1985a.) *A New Vision for Thamesdown.* A Consultative Document. Civic Offices. Swindon.

Thamesdown Borough Council (1985b.) *BREL Works, Swindon. Economic Consequences of Closure.* Corporate Planning Unit, Civic Offices, Swindon.

Thamesdown Borough Council (1991) *Secret of Success: Swindon's Economic Development.* Economic Development Team, Civic Offices, Swindon.

Thamesdown Borough Council (1994) *Thamesdown Trends: Swindon City for the Twentyfirst Century.* Corporate Policy Research Unit, Civic Offices, Swindon.

Thamesdown Borough Council (1995) *Thamesdown Borough Council Development Plan 1996/1009.* Economic Development Team, Civic Offices, Swindon.

Trevelyan, G. (1979) *English Social History, A Survey of Six Centuries, Chaucer to Queen Victoria.* Pelican.

Turnock, D. (1980) *Railways in the British Isles. Landscapes, Land Use and Society.* A & C Black.

Turton, B. (1969) The railway towns of southern England. *Transport History,* 2(2), 105-130.

Vaughan, A. (1977) *A Pictorial Record of Great Western Architecture.* Oxford Publishing Company.

Wells, H. (1950) Swindon in the nineteenth and twentieth centuries. *Studies in the History of Swindon.* Swindon.

22 Violence and the urban environment - reform and development of the housing stock

D. C. DOUGHTY

On the night of July 5th 1995, the unremarkable town of Luton, north of London, burst into the news headlines with one of the worst riots on a housing estate in recent years (Millar 1995; Durham & Tredre 1995). Inner city violence in Britain had usually been confined to outbreaks of tension within the run down and impoverished areas of major cities such as London, Liverpool, Bristol and Birmingham, often related to racial tension and to the factors of boredom arising from unemployment and the exclusive nature of the "enterprise" culture which found little opportunity for the new, underprivileged urban poor.

Britain is by no means alone in its outbreaks of inner city violence, rioting is almost commonplace in certain areas of Paris, Berlin or Los Angeles with causes such as racial tension within areas with high Arab immigrants in Paris, neo-fascist attacks on Turks, Jews and homosexuals in Berlin and black racial tension in Los Angeles. Political violence too has become commonplace with bomb attacks on seemingly non-involved passers by taking place in the hearts of Belfast, Paris, Madrid and Moscow although such attacks may now be on the wane in Britain after the "cessation" of northern Irish sectarian violence.

Inner city violence is thus a mixed bag of causes and effects, be they political, social, criminal or just plain violence for the sake of it. The range of such activities may be organised like the activities of neo-fascist groups at football matches or merely spontaneous like demonstrations in London against government policies such as the poll tax riots or the recent disturbances in Luton not to mention purely irrational attacks on minorities, be they racial or sexist. Nevertheless, many of these disturbances have a certain amount of orchestrated trouble from politically interested extremists

266

as well as the sheer force against society and the common enemy of one's neighbour that mark out both the downside of human nature and the presence of the social misfit or hooligan.

Violence and troublemaking within the city are often considered to be post war phenomena but urban hooliganism or troublemaking under another name is as ancient as the city itself. Ancient Greece and Rome had their riots and insurrections, their violence against the individual both organised by the state and by street gangs and individuals. Medieval cities were terrorised by violence, flourishing states like Venice were plagued by racial attacks, in this case against the otherwise stable Jewish population of the Ghetto (Doughty 1997; Sennett 1994). The Victorian age, which some modern Tories and their supporters in 1990s Britain wish to see revived, gave birth to the term hooligan popularised in a music hall song (Pearson 1983)

> Oh, the Hooligans! Oh, the Hooligans!
> Always on the riot,
> Cannot keep them quiet,
> Oh the Hooligans! Oh, the Hooligans!
> They are the boys
> To make a noise
> In our backyard.   (Music Hall & Theatre Review 1898)

Perhaps the major difference in modern violence within the urban limits and the hooligans of 1898 is the sheer scope and the intensity of the violent attacks of the 1990s. At the same time that the Victorians were satirising their street violence in music hall songs, the neglected Scottish poet, James Thomson was writing his own views on the squalor of Victorian cities, the legacy of which has survived two world wars to be with present day society in the inner city areas of the British metropolis:

> From drinking fiery poison in a den
> Crowded with tawdry girls and squalid men
> Who hoarsely laugh and curse and brawl and fight:
> I wake from daydreams to this real night.   (Thomson 1874)

Victorian Britain saw a vast increase in industrialisation and the growth of the urban sprawl. The mid-nineteenth century saw cities such as

Manchester become simultaneously showplaces of new technology and sites of a previously unknown urban deprivation. It was this period of supposed prosperity and expansion which produced the appalling living conditions that inspired Engels to write his critiques of the state of the British working classes and which gave rise to the street crimes of the time. For the first time, cities had grown to a size and concentration where violence was becoming widespread. The term hooligan was coined after the name of an Irish family and indeed Irish immigration was seen to be a cause of social unrest, drunkenness and a clash of culture with the local inhabitants, just as post-war Britain was to use colonial immigration, that is black immigration, as an excuse for unrest in the 1950s (Pearson 1983).

The very un-British condition of urban violence has thus been on the city streets in one way or another for many years, yet efforts have been made by politicians from Moseley to Enoch Powell to the Thatcherites to blame the condition on something new and often something imported. Vandalism and urban violence are a new sickness caused by black immigrants in the Notting Hill riots of 1959 or the more recent disturbances in Brixton, Moss Side or Toxteth. The permissive society of the 1960s and 70s has been blamed for an increase in lawlessness in the 80s and 90s and economic decline and recession has been given as another creator of the "grab it by any means" society of the present day. Finally, decline in housing stock within the inner cities and the creation of no-go areas run by local Mafias and too dangerous to even police effectively or at all has given rise to gang warfares within and without the areas.

The search for a non-existent "Golden Age" that Nicholas Campion describes in his book, "The Great Year", has exposed such theories that everything was better before the war or in Victorian times or in the Merrie England of the Elizabethan Age. This irresponsible search for hope in a past that could produce fascist riots, Jack the Ripper or the brawling and murders in ale houses such as resulted in the death of the Elizabethan playwright Christopher Marlowe, even led to the summer hysteria of parents fearing for the safety of their children this year (1995). Suddenly, after the murder of two boys and a young girl within hours of each other, children were to be kept close to their parents at all times (Guardian Editorial 1995). Although such murders and indeed the earlier Bulger case where a young child was brutally tortured and murdered by other children, naturally cause fear and resentment, Home Office statistics (Table 22.1) prove that murders of children remain fairly constant:

Table 22.1 Children murdered by strangers (1984-1993)

Age under 5:	15
Age 5-9:	15
Age 10-14:	27

Source: Weale (1995a)

and that this stability in statistics (apart from one unexplained sudden rise in 1974) is contrary to the trend in adult homicides which have increased by over 40 per cent in the last twenty years. Further to this, most of the children murdered are not killed by strangers but by their own parents - in 1993 out of 73 child killings, fifty were killed by their parents (Guardian Editorial 1995). Recent home office reports also suggest that crime is on the decrease, although such statistics should be treated with a certain amount of caution.

In ancient Rome, the ever more violent games of the arena gave a sort of steam valve for the emotions of the public which by the twentieth century had become the working class spectacle of football. Football hooliganism has now been infiltrated by "political" right wing groups intent on bringing chaos and violence into the game, both in the stadium and, even more, in the street of the city before and after the matches. Bill Bufford describes the foolish, over spirited and drunken football fan, living off the fortunes and misfortunes of others but also shows the other side of football hooliganism. His book follows the Manchester United team on various away matches including a game against Turin in Italy where drunken fans frighten the locals before the match but, more worryingly, far right "supporters" wreak havoc with mindless violence on the home supporters and on innocent bystanders. The supposed purpose of the violence is again racist, showing the foreigner who is boss and mixing the force of the mob against the individual for the sake of "taking the town". Football hooliganism is able to wreak havoc in the form of destruction of property, such as throwing dustbins into shop windows, looting damaged premises and much more insidiously, beating up passers by. The football hooligan has developed from the lager lout into a crazed maniac out for blood and havoc and always ready for the situation to "go off" (Bufford 1991).

Undoubtedly this sort of urban violence - for football hooligans always select inner city areas for their pre- and post-match mayhem - relies not only on organised groups, but very much on the effects of alcohol and possibly drugs. Drugs and alcohol are the two factors which have changed the seriousness of inner city violence in developed countries (McKay 1995a). Although the original "Irish" hooligans of the Victorian era created most of their weekend fights as a direct result of too much beer, the availability of "leisure" drugs and the relative cheapness of alcohol, mixed with a society where unemployment is rife, has increased the potential for weekend violence.

Drug dealing, unemployment and the "no future" syndrome, combined with life threatening social diseases such as AIDS has created a youth culture in New York that is portrayed in the new shock film "Kids". The film has caused concern on both sides of the Atlantic for its portrayal of a real life violent society where young teenagers rape, rob and attack each other and the general public with total impunity, where deflowering a young girl by a fifteen year old boy who is suffering from AIDS is taken as natural, acceptable vengeance on society (Abramsky 1995).

The newly emerged states of the old communist block also suffer from a new upsurge in violence. Old time, state controlled violence against its own people, the political enemy or dissident is now replaced by mafia politics and an influx into the cities of gypsies and rural poor who also want their share of the supposed new prosperity. Not that gypsies in St. Petersburg, described by Fallowel in his travelogue of the city with some humour, are only to be found in the new Russia. Rome is plagued by young gangs of gypsy girls, teenagers carrying babies who prey on foreign tourists, picking their pockets as they beg for pennies. Little or nothing is done here as the gypsies belong to that long hated section of society - the foreigner. Like the blacks in Britain or the inhabitants of the old Yugoslavia, they are chased away from any settled homes by the locals. In Birmingham recently, such racial hatred resulted in a black man on the Wyrley Birch Estate being forced to jump 25 feet from his balcony to escape white local attackers and then, whilst lying on the ground with the pain of a broken leg was systematically beaten on the head with an iron bar (Burrell 1995). With such hatred alive between racial groups in the so-called civilised countries the activities of the Roman gypsy girls or the St. Petersburg ragamuffins almost become comical.

Poverty and affluence have always been causes of social and economic tension and the advent of a Thatcherite culture in Britain and similar problems associated with a world economic recession has been a significant cause in the formation of a new greed culture. The "haves" increased their range of possessions in the economic boom period whilst rising unemployment and poverty for the "have nots" led to a philosophy of taking from the rich when the poor could no longer afford to try and keep up. If the neighbours had a video, CD player and personal computer then "why shouldn't I?" led naturally to an increase in theft and aggravated burglary. Friday night pay night persuaded the violent section of the poor to consider it their right to have some of the pickings too and the increases in street crime and muggings may be related directly to a widening of the poverty gap.

Decline in a social system, which arguably was never able to support those it aimed to care for, has led not only to unemployment but to other evils such as a policy of "care in the community" which has unleashed the mentally ill and needy into a community which is unwilling and unable to care. The streets of London, Paris and Hamburg now teem with beggars and mentally unstable casualties of modern times. Not only do major cities house the physically, mentally and sexually abused, but these members of a "no hope" generation and society are forced into prostitution and street crime, threatening the already unstable society of the 90s. Poverty and disease, lack of good housing, the disputable sell-off of better quality state housing, the threats of millennium despair, ever more violent television and cinema (Culf 1995) and hopeless conditions such as AIDS amongst the young all may now produce a lack of trust in the environment and the future of the city.

The social hopes of "homes for heroes" created after 1945 have become ghettoes where violence and discontent, robberies, rape and muggings thrive. Increases in organised and "one off" violence in and around housing estates left to the poorer elements when the better properties had already been bought up by the new aspiring of Thatcher's Britain have led to new phenomena perhaps unthinkable before. The growth of ever more violent female gangs such as the Brentford estate gang who prey on the West End rich and famous as in the case of the attack on the model Elizabeth Hurley (Weale 1995b; Mills 1995). Racial segregation too has led to

271

nationalist gangs on estates heavily populated by ethnic minorities who prey on other groups and sometimes even other sections of their own group (Chaudhury 1995). The housing estate has become a hotbed for social unrest. Falling property prices too have meant that the new homeowners of the Thatcher promise are rapidly becoming the new dispossessed. Le Corbusier's cities in the sky have become hell on earth.

Not every city has fallen this way and if Campion and Sudjic (1994 & 1995) are to be believed, their circular if not spiral effect of history may mean that everything has a chance of change and revival. Garden cities have thrived on their environmental positivism - Brisbane, Australia is one such example of present success whilst Los Angeles, USA is another example of the effect in severe decline. Indeed some areas in the USA are now being left derelict and fenced off whilst richer areas are protected by security gates to keep the well-off in and the others out.

The environmental effect on housing estates and thus on rates of crime is a proposition that was made by experts such as Professor Alice Coleman of Kings College, London whose work on the Mozart Estate in Paddington persuaded the Thatcher government to begin a programme which has spent some £43 million on improvement of seven council estates in Britain since 1988 and which is now under investigation by the firm of Price Waterhouse (Meikle 1995).

It is unlikely to be argued that a pleasant environment, an estate with trees, stripped of the impersonal concrete walkways that join some units to others, less the graffiti and filth in the lift should help promote a feeling of relative well being in the residents.

Handsworth Estate in Birmingham suffered from serious rioting in 1985, an area that had fallen prey to the economic deprivation that was the side effect of Thatcherite boom. Little here has changed; male unemployment stands at 43 per cent (black males at ca. 70 per cent) despite rebuilding and the estate remains a potential powder keg for future rioting (Dodd 1995).

The Mozart Estate in London should be an example of how inner city housing can be changed for the better and how inner city violence and crime can be lessened by environmental improvements. Yet, revamps, removal of concrete walkways, security cameras and better policing may not be the answer that they were hoped to be. The Mozart's central pub was built below residential flats but is now under demolition, the rebuilding programme will also demolish some of the larger blocks and create more individual garden flats, something which should, supposedly, solve the problems associated with vandalism of public green spaces. Yet, despite these improvements, many already in situ, residents still feel depressed by their surroundings - only 42 per cent approved of the removal of the walkways and the crime does still continue (Bowcott 1995). The author's own visit to the estate was prefaced merely hours before by the vandalism of a children's play area - a commonplace occurrence according to the caretakers clearing up yet another mess.

Policing by state and private authorities has also been introduced to the estates. Injunctions on individual tenants to keep down noise, to stop anti-social behaviour and criminal misdeed has proved a success on some estates, notably the infamous Kingsmead Estate in Hackney (Hugill 1995). Surveillance is now wide scale in areas in Newcastle and community action programmes have had some success in problem areas of Wolverhampton. Housing associations are taking over from local authorities but the problem remains that despite all these improvements "the houses still contain the same people" (Simmons 1995).

Sustainable development, within the inner city, requires the efforts of the environmentalists, the planners and the politicians. It also demands engagement from the residents, the people who will or will not benefit from the changes made. Gentrification of housing estates such as the Doddington in Battersea might mean fresh coats of paint, designer panels or prefabricated mock Georgian front doors but often, just around the corner the bored, unemployed youth is fed with cheap drugs and alcohol, torches the derelict houses, ramraids the corner store or sets fire to the rubbish collection. Manchester may have grand schemes to plant forests way into the city centre for the millennium but the sustainability of any city development may still have to rely rather heavily on the sadly lacking education of the country's inhabitants - confection and decoration may well not be enough to stem the tide of riots and discontent that will make any debate on the sustainability of our inner cities no more than a lamentable academic exercise.

References

Abramsky, S. (1995). Where anything goes. *The Independent on Sunday - Travel*. 17th September, p. 57.

Bowcott, O. (1995). Mozart theories fall flat. *The Guardian*. 30th June 1995, p. 7.

Bufford, B. (1991). *Among the Thugs*. London, Secker & Warburg.

Burrell, I. (1995). Racist gangs drive out black families. *The Sunday Times*. 17th September, p. 7.

Campbell, D. (1995). Street-fighting men. *The Guardian*. 19th September, pp. 2-3.

Campion, N. (1994). *The Great Year*. London, Penguin Books.

Carpenter, E. (1883). *Towards Democracy*. London, Unwin Brothers (1926 reprint).

Chaudhury, V. (1995). Enter the rajamuffins. *The Guardian 2*. 19th September, pp. 4-5.

Culf, A. (1995). TV is scapegoat for crime and violence. *The Guardian*. 1st August, p. 8.

Dodd, V. (1995). Life's far from a carnival 10 years after riots. *The Observer*. 10th September, p. 12.

Doughty, D.C. (1997). *Land use policy as politico-racial tool: the case of the Venetian Ghetto*. Paper presented to the 25th Symposium of the European Faculty of Land Use and Development, Szekesfehervar, Hungary.

Durham, M. & Tredre, R. (1995). Boredom rules among Luton's rebel youth. *The Observer*. 9th July 1995, p. 2.

Fallowell, D. (1994). *One Hot Summer in St. Petersburg*. London, Jonathan Cape.

Guardian Editorial (1995). Murders most foul. *The Guardian*. 1st August, p. 12.

Hugill, B. (1995). Estate peacekeepers find ultimate weapon. *The Observer*. 30th July, 1995 p.5.

Lemert, E. M. (1964). Deviance & Social Control, in: Worsley, P. (1970) (ed.) *Modern Sociology*. London, Penguin Books.

McKay, R. (1995). On the mainline to death in world's needle capital. *The Observer*. 10th September, 1995 p. 14.

McKay, R. (1995a). A clean-street recipe for the now mean city. *The Observer*. 24th September, p. 26.

McRae, H. (1994). *The World in 2020 - Power, Culture & Prosperity: A Vision of the Future*. London, Harper Collins.

McVicar, J. (1995). This sporting life of crime. *The Guardian*. 19th September pp. 6-7.

Meikle, J. (1995). Estate revamps fail to cut crime *The Guardian*. 30th June 1995, p. 7.

Millar, P. (1995). Billy the Kid. *The Sunday Times Magazine*. 10th September 1995, pp. 20-26.

Mills, E. (1995). Gentle sex indulges in thrill-seeking violence. *The Observer*. 24th September, p. 7.

Nijkamp, P. & Perrels, A. (1994). *Sustainable Cities in Europe*. London, Earthscan Publications.

Pasolini, P.P. (1955). *The Ragazzi*. English translation by Capouya, E. (1986), London, Grafton Books.

Pearman, H. (1995). When Surrey takes over the country. *The Sunday Times*. 2nd July 1995, p.7.

Pearson, G. (1983). *Hooligan - A History of Respectable Fears*. London, Macmillan.

Peel, L., Powell, P. & Garrett, A. (1989). *20th Century Architecture*. London, Apple Press.

Rook, C. (1899). *The Hooligan Nights*. London, Grant Richards.

Sennett, R. (1994). *Flesh and Stone*. London, Faber & Faber.

Simmons, M. (1995). Tolerate thy neighbour. *The Guardian*. 20th September p.7.

Sudjic, D. (1995). Will it be the dream city we all want? *The Guardian - Context*. 9th September 1995, p. 29.

Sweeney (1995). Hit men are back and the price is right. *The Observer*. September 24th, p. 26.

Thompson, T. (1995). Yardies: myth and reality. *The Guardian 2*. 19th September, p. 5.

Thomson, J. (1880). *The City of Dreadful Night*. Edinburgh, Reeves & Turner & Dobell (this edition as a Canongate Classic, 1993).

Tsoskounoglou, L. (1994). Spatial vulnerability to crime in the design of housing. *The Urban Experience - Abstracts*. Manchester, IAPS.

Weale, S. (1995). Parents' protection dilemma. *The Guardian*. 1st August, p. 2.

Weale, S. (1995). Girlz 'n' the Hood. *The Guardian 2*. 19th September, pp. 6-7.

23 Sustainable development in the Thames Gateway

R. HOME

Introduction

East London is experiencing a major restructuring and remaking of its urban landscape, a process which began with post-war reconstruction, continued with the London Docklands regeneration project in the 1980s and now is marshalled under the strategic concept of the Thames Gateway (formerly the East Thames Corridor). Much of the development land available in the Thames Gateway is either contaminated by past industrial and other processes, or is environmentally sensitive. The estimated 4,000 hectares of development land which have been identified are concentrated on a few large sites, some contaminated by previous uses, and some located by the riverside, including the Royal Docks, Barking Reach, Rainham Marshes (Havering Riverside) and Thurrock. Developers' strategies for mitigating environmental effects and gaining support for the sustainability of their proposals are examined below.

The University of East London, with its campuses at Stratford and Barking, is located in the middle of a major restructuring and reorientation of its sub-region. The London Docklands regeneration project began in the 1980s, and was called at one time the largest urban renewal project in Europe); the process is now continuing under the banner of the Thames Gateway (formerly the East Thames Corridor). The University is participating in the regeneration of the area through its planned new waterfront campus beside the Queen Victoria Dock (scheduled at the time of writing to open in 1999).

Such a major remaking of the urban landscape requires the marshalling of arguments to attract investment and community support for a new vision of the future. Marketing 'keywords' can offer rhetorical support, a range of available and developable meanings for different constituencies. Planning (formerly a powerful key-word) lost consensual support during the Thatcher era, and other key-words have emerged to compete for influence in the setting of agendas, words such as flag-ship development, regeneration,

276

partnership, heritage - and sustainability (Millichap 1993). This paper will explore how the concept of sustainability has been used to justify major new developments in the Thames Gateway, usually in waterfront locations, which are seen as the only way to restore and improve environmentally damaged large sites.

The Thames Gateway: 'a legacy of environmental degradation'

East London and the lower Thames has been called London's 'back yard'. During the Industrial Revolution, London's eastern side, down-wind and down-river, and relatively free from regulatory controls, became a place for the locating of noxious industries, refineries for processing imported raw materials, and ship-yards. It was also the base of most of London's utilities: coal-fired gas and electricity undertakings (Beckton was, in its day, the largest coal gas works in the world, and Barking the largest electricity station in Europe), water reservoirs in the Lea Valley, and hydraulic pumping stations. East London disposed of London's wastes: a network of intercepting sewers, the first comprehensive drainage scheme for a great city, ran to the Abbey Mills pumping station in Stratford, and from there to the Barking treatment works by the Thames. Barges carried refuse to tip sites down-river, and abattoirs, scrapyards and other waste reprocessing enterprises abounded.

Technological change has swept away many of these activities within the last half century, leaving a legacy of derelict and contaminated land, and an under-developed and deprived zone within the affluent south-east of England. East London now faces problems of de-industrialisation, lack of skills for the information technology economy, environmental degradation and physical access, but also offers some of the largest development sites in the region. After centuries of development westwards, London's future growth is slowly concentrating east of the City, starting with London's docklands and spreading along both banks of the River Thames. The London Docklands for a decade dominated the development scene in East London (Brownill 1990), and has now become incorporated into the strategic planning concept of the East Thames Corridor (ETC), devised by the London and South East Regional Planning Conference (Home *et al*, 1992). In 1994, the ETC concept was rebadged and relaunched by government as the Thames Gateway, concentrating on the area between London and the M25, where most of the development opportunities were.

The early studies of the ETC were concerned with its potential for industrial and commercial development, rather than environmental objectives, until the issue of sustainability rose up the policy agenda in the 1990s. Some of the structural choices associated with sustainability were explored in a 1993 key report by planning consultants (Llewelyn-Davies 1993) on the development capacity and potential of the ETC. Among them were a number of transport issues: reducing the need to travel through mixed use activity on major sites, reducing reliance on long-distance commuting, organising the development structure around public transport, and getting better use out of the existing road space. When regional planning guidance was published (RPG9A, 1996), it claimed to be setting 'the framework for a sustained and sustainable programme of economic, social and environmental regeneration'. Fourth in the five objectives defined for the area was:

> to encourage a sustainable pattern of development, optimising the use of existing and proposed infrastructure and making the fullest possible use of the many vacant, derelict and under-used sites which previously supported other activities.

The document went on to claim that 'it is the scope for meeting development needs whilst also achieving environmental objectives that distinguishes the Thames Gateway opportunity'. So the conversion of derelict and contaminated sites into new development was justified by environmental improvements which could be incorporated in the form of planning gain. The Thames Gateway also offered the opportunity to direct new development away from the attractive landscapes elsewhere in the region

> It is unlikely to be a coincidence that Michael Heseltine's original announcement (declaring his vision of the ETC in 1991 as Secretary of State) came after five politically bruising years when politicians in counties to the west and north of London were in open conflict with themselves and the DoE over the resistance of existing residents to large-scale residential construction (Church and Frost 1995: 208).

As one of the authors of the Llewelyn-Davies study pointed out in 1994, the Thames Gateway 'offers the only part of the Region where land needs -

primarily for housing - can be met without major, expensive, delaying and politically damaging controversy' (quoted in Church and Frost 1995).

The major development sites: seeking 'a new environmental standard'

The Thames Gateway planning framework saw environmental improvement and economic regeneration complementing each other 'to break out of the self-reinforcing cycle of environmental blight' (RPG9A 1996, paragraph 4.4). The image problems of the area created the call for 'a new environmental standard'

> There must be confidence that there is a commitment to improving the environment and to sustaining that environment. It is essential that the decisions and actions of local authorities, and other interests - both public and private - work with the grain of this expectation of environmental improvement.

Within the core area, an estimated 4000 hectares of development land are concentrated on a few large sites, some contaminated by previous uses, mostly with the advantage of a riverside location, and most needing transport investment.

It is the installation, over a 20-30 year period, of major new transport infrastructure, that is central to the regeneration strategy. An estimated £2 billion of capital spending on roads and £2.5 billion of rail projects is committed through the 1990s. Although the Department of Transport stated in 1989 that 'the Government does not intend to build urban motorways to encourage more car commuting into Central London', both radial and orbital road capacity has been significantly increased. Within the last decade, the following have been completed: London's orbital motorway (the M25) with its high-capacity river crossing at Dartford, an upgraded North Circular Road, and the Docklands road network. Over the next decade further key components will be installed: the A13/M25 links, and the M11 extension into Hackney. Investment in other modes of transport include the Docklands Light Railway, the Stansted passenger terminal and Docklands City Airport, the Jubilee Line extension to Stratford, and the Channel Tunnel fast rail link, re-routed through East London to a terminus at St.Pancras, with intermediate stations at Ashford, Ebbsfleet and Stratford.

With development opportunities in the Thames Gateway located in a few sites, each local authority has been promoting its own special project, in a striking example of what Michael Hebbert (1991) has called 'the borough effect' upon London's geography and physical development. The larger authorities, created by local government reorganisations in 1965 and 1974, compete with each other to attract development. Moving from the City down the northern bank of the Thames, one encounters a succession of major riverside sites, each being actively promoted by its local authority as a development opportunity (Thames Gateway London 1996).

First is the Royal Docks and Beckton Gasworks, in the London Borough of Newham. With new construction largely complete in the LDDC's western areas (Wapping, the Isle of Dogs and the Surrey Docks), attention has moved to the Royal Docks, east of the Lea River and representing one of the single largest redevelopment sites in Europe (194 ha./479 acres of land and 96 ha/237 acres of water). The transport infrastructure for the Royals is largely complete, but development has been hard to attract. Landscaping and creative use of the enclosed water of the docks are important priorities, while the huge Beckton Gas Works site at Gallions Reach (60 ha/150 acres) requires special decontamination measures.

A few hundred metres downstream, in the adjacent borough of Barking and Dagenham, across Barking Creek, lies the Barking Reach development site (250 ha/600 acres, in several ownerships). Previously the Barking Levels and Dagenham Marshes, and interspersed with numerous industrial enterprises, this somewhat desolate tract has now been allocated for major residential and employment development, almost on a new town scale. A framework plan prepared in 1988 envisaged 'housing-led development, which would encompass the full range of community facilities, as well as new commercial, business and leisure uses'. Central government and European Union assistance is funding some of the decontamination and environmental improvement measures needed, such as the importation of landfill material and the construction of a canal.

Moving further downriver, past the Ford motor works, one comes to the Rainham Marshes, rebadged by the London Borough of Havering for marketing purposes as 'Havering Riverside'. Covering some 1600 acres/650 hectares and 3 miles of river frontage, the site includes silt lagoons, created by dredging from the Thames, and extensive ground contamination, while part of the site lies within the Inner Thames Marshes Site of Special Scientific Interest (SSSI). While these are major environmental constraints, Serplan conceded that part of the SSSI could be lost for economic development, perhaps incorporating an 'ecological park' into any new development. Attempts to capture a station for the Channel Tunnel rail link (which will pass the site) were unsuccessful, but have not deterred the borough's promotional efforts. Various development proposals have come and gone, including a major leisure scheme and an international exhibition centre. A 1990 Serplan report suggested that proposals for the site should be

> sufficient in scale to create a new market identity and to bring perceptions held by investors and developers up to a level to match areas in the west of the region.

Leaving the boundaries of the London boroughs, one enters the Essex district of Thurrock, which has to date been highly successful in attracting development to its riverside areas east of the M25 crossing. A long-term strategy of using development to restore the vast tracts of worked-out chalk pits in the area (Essex 1976) has resulted in the Thurrock retail park, and the Lakeside regional shopping centre (the largest out-of-town shopping centre in the south of England). Across the railway line at Chafford Hundred, hundreds of hectares of disused chalk quarries were granted planning permission in 1987 for five thousand houses, shopping and community facilities, in what has been called the UK's largest current private new settlement. Years of negotiations between Essex County Council, Thurrock Borough Council, and the various development consortia were much concerned with environmental improvements, to justify releasing the derelict land for development. Marketing for the Chafford Hundred housing scheme emphasised the ecological issues, making the flooded chalk-pits into environmental assets, their cliffs supporting sandmartins and bats, and picnic areas, nature reserves and information points projected in 'Chafford Hundred's heart of green'. The developer claimed to have rerouted a main road to allow a rare spider, known only in the Grays area, to remain in its natural habitat (Chafford Hundred 1996).

A similar pattern, of large reclaimed or decontaminated sites vigorously promoted by their local authorities for development, can be traced south of the river, from the Greenwich Peninsula site (reserved for the

281

'Millennium Experience' in the year 2000, with central government backing) to the riverside towns of northern Kent. The proposals have all marshalled arguments linked to sustainable development and environmental protection.

The River Thames: 'London's unsung asset'

Running through the Thames Gateway is, of course, the river itself; a barrier to be crossed, but also an opportunity for sustainable development approaches. The removal of much of its commercial shipping, and the growing strength of the environmental movement, has resulted in a new view of the river, with land use planners having the opportunity to become involved in managing competing land uses, rather than being excluded by the Port of London Authority, with its primarily commercial brief.

A succession of policy documents from different public agencies have redefined the view of the Thames over the past two decades, reflecting the growth of land use planning and the decline of the docks. In 1974, the Greater London Council organised a conference about the lower Thames, and in 1985 produced a 'Manifesto for the Use of the River'. In 1987, the London Planning Advisory Committee created a Thameside Working Party, comprising 37 organisations, which produced a set of guidelines. A London Rivers Association was created, seeking to ensure the integration of London's rivers into the Capital's urban and social fabric through the maximisation of their economic, transport, amenity and ecological potential. When in 1993, Secretary of State Gummer initiated a consultation on the future of London, nearly half of the respondents said that the River Thames was one of the features that they most appreciated about London. The government published, in the following year, a consultant's study by Ove Arup, called the Thames Strategy (GOL 1995). Gummer, in his forward to it, referred to the river as 'London's great unsung asset', and spoke of 'my quest to challenge perceptions of the river', restoring it

> as a unifying element, a welcome break in the city's unremitting urban form, rather than perpetuating an outmoded notion of the river as a physical and psychological barrier dividing the capital.

Ove Arup saw the Thames as fulfilling at least six major functions: as a major linear open space, a recreation and leisure facility, a transport artery, a setting for major buildings, a source of drinking water and means of removing waste, and a conservation and ecological resource.

The Thames Gateway planning framework has also emphasised the waterfront aspect, including in its 'framework principles'

> The economic and environmental potential of the river and riverfront and the need to avoid loss of the waterfront to developments which do not benefit significantly from a riverside location: bringing life to the river and riverfront (RPG9A 1996: 4.3)

This reflected, in part, the PLA's concern at the loss of wharfage facilities, and the need for longterm safeguarding of the limited number of sites suitable for port development (with deep, sheltered water access, flat topography and good land access). Meanwhile, the demand for waterside development has grown, and the LDDC has claimed a commitment to opening up the waterside to public access (although progress has been poor, with private developers regarding restricted access as a significant source of added value) (for waterfront development, see Hoyle *et al* 1987, Breer and Rigby 1994)

While much of the new policy initiatives concentrated upon the Thames within London's boundaries, issues of environment and sustainability have recently focused attention on the lower reaches of the Thames Estuary. Following a House of Commons Environment Select Committee report in 1992, which identified a need for strategic estuary planning and management (British estuaries comprising a quarter of the estuarine resources of Western Europe) (Lee 1993), English Nature launched the Thames Estuary Management Project. The Thames is now far narrower than when the Romans settled on its banks, with vertical walls confining the river down to the marshes of the lower estuary, where the tidal foreshore offers a complex mosaic of ecological habitats, mainly marine shallows and intertidal flats, saltmarshes and coastal grazing marshes, which provide one of the five most important winter feeding areas in Europe for wading birds. Large areas are designated as Special Protection Areas, Sites of Special Scientific Interest, and international wetland bird breeding habitats, even including some of the derelict and contaminated land (such as the Rainham and Dartford marshes) (EA 1996).

The Thames Estuary Management Project (TEMP) seeks to raise the profile of the estuary, which it defines as reaching from Tower Bridge downstream on both banks, to Shoeburyness on the north and the Isle of Grain on the south. A process of identifying 'key players' (comprising landowners, local authorities, user groups and others) was followed by each being invited to prepare a 'statement of interest'. The Thames was revealed as Britain's busiest and most socially complex estuary by the diversity of the issues agreed for individual treatment in the TEMP: archaeology, fisheries, flood defence, nature conservation, publicity & information, recreation & access, the port, the need for strategic management, the Thames as a transport artery, waste management, water quality & pollution, and waterside development (Kennedy 1994 and 1995).

Conclusions

This paper has shown how the half-completed London docklands development has been extended to the wider sub-regional context through the informal planning mechanism of the Thames Gateway, deploying arguments of sustainable development, to reflect the growth of the environmental movement in the 1990s, and to achieve funding for transport infrastructure improvements and land decontamination. Various rhetorical terms were incorporated into the argument, including raising the poor image of the area and improving the environmental standard. For the Thames estuary itself a new informal machinery of the estuary management plan sought to bring together the main 'stakeholders' and overcome, or at least reduce, some of the complex land use conflicts, while throughout the area local authorities competed with each other to attract reclamation funding and development interest. There has been little attempt by central government, committed under the Conservative administration to free-market, non-interventionist approaches, to direct investment and choose between the many competing projects. Indeed, a major role of the Thames Gateway has been to provide a planned alternative to lessen some of the development pressures elsewhere in the environmentally superior areas of the outer South-East, while also contributing towards a environmental clean-up of the vast tracts of riverside land despoiled by past industrialisation.

References

Astbury, A.K. 1980. *Estuary: Land and Water in the Lower Thames Basin*, Carnforth Press, London.

Breer, A., & Rigby, D. 1994. *Waterfronts*, McGraw-Hill, New York.

Brownill, S. 1990. *Developing London's Docklands: Another Great Planning Disaster*, Paul Chapman, London.

Chafford Hundred 1996. Promotional literature by Chafford Hundred development consortium.

Church, A., & Frost, M. 1995. The Thames Gateway - an analysis of the emergence of a sub-regional regeneration perspective. *Geographical Journal*, **161**, 199-209.

EA (Environment Agency) 1996. *The Tidal Foreshore*, Environment Agency, London (in partnership with London Ecology Unit).

Essex, J. 1976. Tackling land dereliction in Thurrock. *Papers of the Land Reclamation Conference*, Thurrock Borough Council, 46-84.

GOL (Government Regional Office for London), 1995. *Thames Strategy*, Dept of Environment, London.

Hebbert, M. 1991. The borough effect in London's geography. In: *London: a New Metropolitan Geography*, (edited by K.Hoggart & D.Green), 191-206. Edward Arnold, London.

Home, R.K. *et al.* 1992. *Business Activity and Land Use Patterns in the North East Thames Corridor*. Department of Estate Management, University of East London.

Hoyle, B.S., Pinder, D.A., & Husain, M.S. 1987. *Revitalising the Waterfront*, Belhaven, London.

Kennedy, K. 1994. *Issues on the North Thames - A Case for Estuary Management*, English Nature, London.

Kennedy, K.H. 1995. Producing management plans for major estuaries - the need for a systematic approach: a case study of the Thames Estuary. In: *Directions in European Coastal Management* (edited by Healey, M.G., and Doody, J.P), Samara Publishing: Cardigan, Dyfed, 451-9.

Lee, E. 1993. The political ecology of coastal planning and management in England and Wales: policy responses to the implications of sea level rise. *Geographical Journal*, **159**, 169-78.

Llewelyn-Davies and Department of the Environment, 1993. *East Thames Corridor: A Study of Development Capacity and Potential*, HMSO, London.

Millichap, D. 1993. Sustainability: a long-established concern of planning. *Journal of Planning & Environmental Law*, 1111-19.

NRA (National Rivers Authority), 1994. *Thames 21 - A Planning Perspective and a Sustainable Strategy for the Thames Region*, Consultation Draft, London.

RPG9A 1996. *Thames Gateway Planning Framework*, Department of Environment, London.

Thames Gateway London, 1996. *Development Opportunities in the Thames Gateway*.

24 Reclamation of abandoned land leading to sustainable development

R.K. BULLARD

Introduction

There is an assumption in the popular press that sustainability may only be concerned with rural areas and an even wider assumption, that it might only be appropriate for developing countries, and that the countries of Europe may not be so concerned with this topical activity. The need for implementing sustainability is a world-wide requirement, and one which should not be limited to any one country or region. As Europe and the other developed countries of the world are the major consumers of it's resources, it follows that they should be at the forefront of the quest to achieve sustainability.

Numerous international agencies, non governmental organisations (NGOs), and World Conferences have used the word 'sustainability', but there is, to date, no clear or internationally accepted definition, except possibly that of the World Commission on Environment and Development

> development that meets the needs of the present without compromising the ability of future generations to meet their own needs (Habr 1995).

The other well-known definition is

> a process of change in which the exploitation of resources, the direction of investments, the orientation of technological development and institutional changes are made consistent with future as well as present needs (Bruntland 1987).

In the built environment there will be a need to ensure that buildings are recycled, that the materials from which they are built can be used again where the building itself cannot be refurbished. With an increasing demand

287

for aggregate the crushing of waste concrete to provide sand and gravel this is a viable example of an attempt to achieve sustainability.

In rural areas the coppicing of trees, a practice that had virtually died out through lack of use and expertise, is now becoming more acceptable to the extent that young people are being trained in the techniques. The use of timber that is cut from sustainable forests is becoming ecologically acceptable. Some users of the timber and distributors of the products, advertise and promote this issue in their sales techniques, 'only timber from sustainable forests', with the suggestion that no one in their own interest would use timber or products from other sources. Land formally termed 'set-aside' by the European Commission of the European Union is now being used in England for growing willow trees. These are a fast growing species and the timber, a sustainable resource, is a potential energy source in converted power stations.

A further step beyond the sustainable activity is towards progressive development where, 'building blocks' for the future are introduced. 'Building and conserving for future generations' should be the slogan for the property profession.

One of the concerns that must confront the population in Europe and especially that of North America, is how sustainability can be achieved when six per cent of the world population is consuming more than thirty per cent of the world's resources. With reference to the opening statement in this chapter, it would seem that it is the developed countries who need to take on the issue of sustainability, especially those living in the urban areas. Many of the developing countries through lack of, or perhaps because of limited resources, are forced to recycle increasing amounts of their or the waste of other's, thus setting an example to the developed countries.

In order to achieve sustainability, a number of extreme measures might have to be undertaken by governments, some of which may be copied from the examples taking place in developing countries. In an attempt to reduce the exploitation and consumption of resources, these extreme (by European standards) measures might have to be taken (Goodland *et al* 1993). These may include some of the following; the need to limit their population growth, the need to limit their affluence, and the need to improve technology.

Examples may be found both at national and at regional levels in both developed and developing countries, some being government policy, or being introduced as the result of economic necessity, with others having been introduced through expediency.

Abandoned land

Land is abandoned for a variety of reasons, not least in the rotational agricultural practice where the land is left fallow. When the land is 'resting' or out of production, it might be taken as being abandoned. This ties in with the definition of 'land not in use'. The time period of abandonment might not be important, though from an ecological point of view the time may be important for natural regeneration to occur. From the monitoring of abandoned land, the land use, as recorded by an airborne sensor, may detect the presence but not necessarily the duration of abandonment (Bullard 1981; Bullard & Lakin 1982). With the need to establish duration this might be more clearly established during a subsequent ground survey.

Abandoned urban land

The abandonment of land in urban areas often occurs as part of the development process. As buildings pass their economic span, they fall into disrepair and disuse. With no or few occupants, this process is speeded up, and while the land awaits redevelopment the land remains abandoned. Nijkamp & Perrels (1994) state that many people consider the urban area as

> concentration points of environmental decay, where pollution, noise annoyance and congestion can mean a serious threat to human welfare and well-being.

This is probably overstating the case, but it does emphasise the 'downward spiral' as opposed to the 'upward spiral' that is desirable for sustainable development.

In parts of London, homes and shops have been compulsorily purchased for proposed new infrastructures and new developments. There is inevitably an interval of years before the schemes are implemented. In extreme circumstances and for a variety of reasons, the development has not taken place, and all that remains is abandoned land surrounded by rusting corrugated iron fences to keep trespassers away. The need for the better use and recycling of land use and buildings in such areas is therefore needed.

This phenomena of abandoned urban land could be brought about by inadequate planning control, which has led to 'urban sprawl'. The 'waste' lands (Haughton & Hunter 1994) may only be reclaimed when a major development takes place. Examples of this waste commonly occur on fast expanding cities, Toronto being an extreme example with an apparent large amount of waste occurring between the areas of development and urban encroachment. Abandonment of land, or non-sustainability, in a city could imply a structural decline of the economic base of a city (Nijkamp & Perrels 1994). This could be the result of some or all of the following; population decline, environmental degradation, inefficient energy systems, loss of employment, relocation of industries and services, and unbalanced social demographic composition. Many of the above factors are familiar in certain European cities, and these usually have the largest areas of recorded derelict land.

Some of the abandoned land is classified as 'brownfield' sites. These consist of derelict land that contains some degree of contamination and therefore requires 'cleaning up' before it can be used for development. A topical site is the Greenwich Peninsular upon which the Millenium Dome is being built. The site was described as

> one of the untidiest and foulest-smelling corners of Britain. (Prentice 1997).

The Greenwich site contained a former British Gas works, which was closed during the 1970s. There will be some 5,000 homes built on the site, which the council conceded will not be clear of contamination when building commences. It is estimated that some 2 million houses will be built on brownfield sites in Britain during the next twenty years.

Abandoned rural land

Abandoned land in rural areas has occurred over the millennia. The extraction of minerals has taken place throughout Europe, and many of the abandoned sites are still classified as derelict (Bullard 1981). The south-west of England has the largest concentration of derelict land but one of the smallest quantities of urban dereliction. The extraction of lead (started before Roman times), copper and coal deposits in this part of England, has almost completely been abandoned, and large areas of waste land and buildings is the legacy of this once valuable resource.

Changes in farming practice have not taken land out of production. The intensification, brought about through the need to be self sufficient, has changed the land use pattern by increasing field size and thereby reducing habitats. The agricultural intensification has also led to environmental decline (Hoggart *et al* 1995).

The introduction of 'set-aside', the taking of land out of production, has given the farming community in particular, and the planners in general, the opportunity to reconsider rural land use. The set-aside policy has allowed selected land in rural areas to be taken out of agricultural production. This process has not yet been specifically related to certain tracts of land for a permanent use as set-aside. It is now increasingly important that decisions are taken on this important issue to allow the land, so set-aside, to be permanent if this is required by any of the parties involved. If there is to be a rotation in the use of the land, this may be beneficial to the farming community and rural areas, but may be damaging to the periphery areas. This conflict must be solved if the benefits of urban greening are to be achieved (Bullard 1996).

Reclamation in the urban context

In Europe, the urban population is, on average, just over seventy per cent of the total. In the UK, this figure increases to over ninety per cent. By comparison, the figure for Africa is about thirty per cent, though this is rapidly increasing, as it is in most developing countries as rural inhabitants flock to the towns and cities.

In the European context, therefore, the need for reclamation leading to sustainability is of more importance in urban areas, because the greatest number of the population who will need to be involved in order to achieve this objective. However, some of the land that is currently abandoned should not be brought back into permanent use (Elkin *et al* 1991). Land that was formerly railway routes and marshalling yards could be needed again when the full environmental costs of road transport is appreciated, and when action is taken to improve public transport and reduce reliance on cars in urban areas. The abandoned routes could also provide a swathe of habitat for wildlife and a corridor along which species can travel in relative safety.

Urban reclamation for development

In the Essex Borough of Thurrock, the proportion of derelict land area was one of the largest in England (Bullard 1981). This borough has now become an attractive area for development because of the construction of the M25 motorway, and the new Queen Elizabeth Bridge over the River Thames at Dartford. The derelict quarries of Thurrock, to the east of the M25, have been transformed into the Lakeside development, the largest shopping centre in Europe with 12,000 parking spaces.

Even with this number of parking spaces, the cars spill out into the surrounding urban area at weekends, causing considerable congestion and unwelcome intrusion into the lives of local inhabitants. The Swedish firm IKEA is building an outlet in the area, subject to a £1 million parking area being provided by the Local Authority. This could ultimately lead to further congestion and conflict of interests. The development is, however, providing employment in an area, which was losing jobs as the main industries and employers in the area (quarrying, cement factories and associated industries) closed down.

Reclamation to improve urban ecology

A new approach to protecting the urban environment is referred to as 'urban ecology', which sets out a number of design principles (Nijkamp & Perrels 1994);

1. minimise space consumption in urban areas (e.g. underground parking areas),
2. minimise spatial mobility in the urban space by reducing the geographical separation between working, living and facility spaces,
3. minimise urban private transport,
4. favour the use of new information technology and telecommunication technology,
5. minimise urban waste and favour recycling,
6. minimise urban energy use (e.g. combined heat and power systems, district heating etc.).

These could lead to multiple land use, and improvements leading to less expenditure of resources and reduced costs.

The restoration of rivers has become an important activity both in the urban and the rural areas (Biggs & Williams 1993). Many rivers have deteriorated in quality and in the process has negatively influenced their surrounding environment. Many rivers have been straightened, deepened, diverted and contained. The river valleys and their floodplains have been intensively farmed or developed and with this, the aesthetic and ecological benefits have been lost.

An urban example of river restoration may be found in the north east of England, at Darlington on the River Skerne. The River Skerne is typical of many rivers in their middle and lower reaches, which are flowing through built-up areas. They are cloudy, contaminated, and slow-moving. The River Skerne carries fine silt that smothers aquatic life. Nettles, oilseed rape, and docks cover the riverbanks. Roads, railways, pipelines, housing, factory development and industrial tipping occupy much of the floodplain. Reclamation of the River Skerne, have led to environmental improvements and, it is suggested sustainability, have included the following;

1. creating new bends in the rivercourse, and sloping banks, making the river safer,
2. allowing new shelves of land to develop on the inside of bends to become filled with reeds, water plants and flowers,
3. naturally strengthening banks with trees and reeds,
4. creating wetlands and backwaters to attract new wildlife and provide havens for fish and dragonflies,
5. modifying sewage outfalls to reduce the risk of pollution and improve the appearance,
6. adding a second footbridge, and extending walkways for public enjoyment,
7. enriching the parkland with native trees, plants and flowers that will attract flora and fauna,
8. involving local people in the new development.

With the above improvements, the residents and visitors to Darlington will ideally reap the benefits, and consequently see their river as an asset rather than as a dumping ground. With a healthy balance between the interests of the residents and achieving ecological sustainability, in the context of facilities for future generation, this now will be possible.

Reclamation in the rural context

As with reclamation in the urban context, rural reclamation will be required for developments, largely identified as agriculture since this is the major rural industry, and also to improve the ecology. Since the Second World War much of the rural land formerly used as meadow (especially water meadow), marshland, and other unused or under used land has been brought into production. Marshes have been drained, rivers have been straightened, and meadows have become arable fields. Field sizes have increased, hedges have been removed and habitat has been lost. Losses in soil through water and wind have increased as the natural defences have been broken down.

In an attempt to increase production, especially after the shortage of food and the demand for self-reliance, the countries of Europe have encouraged farmers to produce more. Although the demand for food is ever present, with special demands in developing countries, there is now a move to reduce production in Europe, and this has given the ecologists and others the opportunity to 'redesign' the rural land use pattern.

Rural reclamation for development and agriculture

Where once there were grants available to farmers for them to increase the size of their fields by the removal of hedge rows and small groups of trees, there are now grants available for land owners to plant hedges and coppices. Priorities are changing, but the population still needs feeding, and self reliance is as important as ever, especially with droughts occurring in many parts of the World and countries, such as China, now increasingly reliant on rice imports. The impacts of global warming and the increased occurrence of *El Niño* are taking their toll on World resources.

Reclamation in the agricultural context, should be seen as an attempt to farm the land by taking into account the ecology, the impact of fertilisers and pesticides on the environment, the returning of waste back into the ground, and the need to consider an environmental audit. The cost of damage to the land should be equally levied in rural areas as they should be in urban areas. The expression, "the polluter pays", should apply equally to land owners and users in rural and urban areas. The reclamation of land for development will have to ensure that new buildings and infrastructure have a less damaging impact on the rural environment than they have had in the past.

Of concern in Britain are the 2.4 million new houses that are scheduled to be built into the next millennium. Many of these houses will have to be on green-field sites, within green belts and in rural areas. Ironically, there are probably not sufficient abandoned or brownfield sites to satisfy the builder's demands for suitable development land.

Reclamation to improve rural ecology

The altering of water courses in the past, by straightening them to increase the flow of rivers, was seen as a way of dispersing the water as quickly as possible. The reclaimed land in the former flood plains, was farmed for the cereal crops of Europe, and led to the 'Grain Mountain' of the European Union. The periodic flooding of the former flood plains had disastrous results on the farmland, and surrounding habitation. Providing that this is not too frequent, it could be accepted as it was in the past. However, flooding is becoming more frequent and the scale of damage is increasing.

The grasslands of reclaimed land need to be self-sustaining and it may no longer be acceptable even to reap hay or silage from the area set aside for the new water meadows. Many of the areas that were previously used for arable farming, require time to regenerate into 'natural' habitats. Below the grasslands, buffer strips can be built which can act as traps to further reduce the flow of the river. By reinstating the original 'winding' river, the flow is reduced and siltation is thereby encouraged.

Reclamation for river restoration

With the change in land use priorities, there is now a positive encouragement to promote environmentally beneficial reclamation. The planned construction of wildlife habitats are examples where changes in land use priority can be implemented within a framework of land consolidation schemes in, for example, the Netherlands.

In Denmark and the UK, there is increasing attention being paid to the benefits of river restoration in rural areas (Jensen 1992: Biggs & Williams 1993). This process is now possible because land, formerly under cultivation, is now being allocated for its former purpose; i.e., to provide a flood plain and/or a water meadow. In so doing, this process will prevent many of the excess chemicals from leaching into water courses (Harper &

Ferguson 1995). River restoration was undertaken in Denmark and the UK with the formation of a European Community LIFE Demonstration Project (River Restoration Project Limited is the organiser of the activity in the UK). The projects have been supported many government, NGOs and voluntary bodies.

Achieving sustainability

The achievement of sustainability will probably be one of the most difficult tasks that any government or agency is likely to tackle. If the definitions quoted above are to be taken literally, all future planning would have to look towards, anticipate the needs, and provide the resources for future generations. This approach would have serious implications on the use of existing resources.

Land must be considered as the ultimate resource for without it life would not exist. Unless the land is made sustainable then any other issue will be of little or no consequence. Factors needed to achieve land sustainability include; multiple land use, diminish the extraction of non-renewable resources, increase the use of renewable resources, apply an environmental audit, make the polluter pay, and provide better planning for the future.

The two major issues on sustainable development that were raised by the United Nations Conference on the Environment and Development (UNCED), at its Rio Conference, which were of importance are as follows;

1. improved mobility management including more efficient and environmentally rational location and transport modes,
2. coherent packages of measures to achieve improvements in environmental quality in urban areas.

For readjustment, in both rural and urban areas, the objective would be to use the land to its full potential, to avoid further dereliction, and to reclaim where appropriate all derelict and contaminated land, thus sustaining the major resource, land (Bullard 1993). The amount of waste (abandoned land) that is caused must be kept below the assimilative capacity of the environment (Van Kooten 1993).

The Dutch National Environmental Policy Plan (NEPP) has three broad, long-term objectives for achieving sustainable development namely; that the consumption of energy than no longer be recovered from the sun, the treatment of all wastes as raw materials, the promotion of 'high-quality' products that last, can be repaired and are suitable for recycling. These three objectives must be strived for, and ideally achieved, if sustainability is to be accomplished.

Data for sustainability

According to Fellman (1994), the current real property and environmental regulatory information systems block the information process required to achieve environmental sustainability. What is required is 'deep information', that is, the integration of local government data at land parcel level. The data will, of course, only be one step in the process of sustainability, though an important one. In addition, acquiring and using this information to the benefit of present and future generations will be of importance.

Conclusions

The bringing of abandoned land back into use will greatly assist mankind in achieving sustainability. One of the concerns will be the interpretation of the word 'use'. In designing future land use activities, more attention must be made in allocating land for wildlife and habitat and for environmental concerns.

The following factors will be of importance in undertaking *urban reclamation* to achieve sustainability;

1. population density of urban areas will necessitate that inhabitants are a party to and support reclamation,
2. reclamation may not be one hundred per cent as derelict land may itself be required in the future,
3. reclamation will be needed to improve urban ecology.

The following factors will be of importance in undertaking *rural reclamation* to achieve sustainability;

1. replanting will be a priority,
2. agricultural land must be environmentally compatible with sustainable objectives,
3. recycling of farm waste,
4. reduce pollution into water course,
5. river restoration.

The following factors must be recognised for *sustainability* to be achieved;

1. land is the ultimate resource,
2. that only that energy recovered from the sun should be consumed,
3. all waste, including that already buried, should be considered as being potential raw materials,
4. data required to assist achieving sustainability has to be at parcel level, enabling in depth analysis.

For the future sustainable requirements, perhaps we need to reconsider our priorities and to learn from those less well off than ourselves. We need to know how to survive without consuming all the earth's resources during our own generation, and how to leave enough for future generations.

It is not Earth that's the problem, it's the lifestyle of western societies. (Lynn Margulis, Member of the US Academy of Sciences).

For the human race to survive, the land on which we all live must be defended from those who wish to destroy the very natural ecosystems that regenerate the planet.

Acknowledgements

Thanks are given to the following for their support and for the provision of materials. However, the text is that of the author, except when others are quoted. To Mr. Alister Driver, NRA Thames Region; Miss Victoria Jones of Pebble Mill Studios, BBC, who enabled other contacts to be made; Mr. Richard Vivash, of the River Restoration Project, and Mr. John Young of the National Trust who first showed the author river restoration on the River

Windrush, on the Sherbourne Estate. Thanks for their assistance and support.

References

Biggs, J. & Williams, P., 1993. *Feasibility Study - River Restoration Practice Throughout Europe*, The River Restoration Project Ltd., Huntingdon, England.

Boon, P.J., Calow, P., & Petts, G.E., (eds.) 1992. *River Conservation and Management*, John Wiley.

Bruntland, G.H., 1987. *Our Common Future - The World Commission on Environment and Development*, Oxford University Press, Oxford.

Bullard, R.K., 1981. *Abandoned Land Detection, Reclamation Proposals, and Monitoring, Aided by Remote Sensing*, unpublished PhD thesis, Faculty of Pure Science, University of Sheffield.

Bullard, R.K., 1990. Environmental Impact of Land Consolidation, *Surveying Science in Finland*, 8(1), 31-36.

Bullard, R.K., 1996. Land Readjustment for Restructuring Urban Dereliction, Paper presented at the, *European Faculty of Land Use and Development, 19th International Symposium on Land Readjustment as an Instrument for Management of Landscape and Urban Green Space.* In, Tayama & Weiss (Eds.) Land Use Problems in the Urban Periphery, Lang, 55-76.

Bullard, R.K., & Lakin, P.J., 1982. Remote Sensing's Role in Monitoring Reclaimed Land, *Chartered Land Surveyor/Chartered Minerals Surveyor*, Spring 1982, 3(3), 4-14.

Elkin, T., McLaren, D., & Hillman M., 1991. *Reviving the City - Towards Sustainable Urban Development*, Friends of the Earth, London, U.K.

Fellman, J., 1994. Deep Information: The Emerging Role of State Land Information Systems in Environmental Sustainability, *URISA, Journal of the Urban and Regional Information Systems Association*, The University of Wisconsin Press, 6(2), 11-24.

Goodland, R.J.A., Daly, H.E., & Serafy S.El, 1993. The Urgent Need for Rapid Transformation to Global Environmental Sustainability, *Environmental Conservation*, 20, 297-309.

Habr, H.El, 1995. Freshwater Resources: their Depletion, Contamination and Management, in: *Environmental Management: Issues and Solutions*, eds Atchia M., and Tropp S., John Wiley & Sons Ltd., Chichester, England, 86-90.

Harper, D.M., & Ferguson, A.J.D., 1995. *The Ecological Basis of River Management*, John Wiley.

Haughton, G., & Hunter, C., 1994, *Sustainable Cities*, Regional Studies Association, Jessica Kingsley Publishers Ltd., London.

Hill, C., 1992. EC Takes Environmental Action for Sustainable Development, *Chartered Surveyor Monthly*, **1**(8), 1.

Hoggart, K., Buller, H., & Black R., 1995. *Rural Europe: Identity and Change*, Arnold, London.

Jensen, J., (ed.) 1992. *Nature Management in Denmark*, The Ministry of the Environment, The National Forest and Nature Agency, Kobenhaven, Denmark.

Nijkamp, P., & Perrels, A., 1994. *Sustainable Cities in Europe*, Earthscan, Earthscan Publications Ltd., London.

Prentice, E.-A. 1997. Peninsular war on poison plot, *The Times*, **66,051**, 43.

Van Kooten, G.C., 1993. *Land Resource Economics and Sustainable Development*, University of British Columbia Press, Vancouver, Canada.

Index